Getting Started
with
Maple®
Second Edition

C-K. Cheung
G. E. Keough
Michael May, S.J.

Boston College and
St. Louis University

WILEY

John Wiley & Sons, Inc.

D1409826

ISBN 0-471-47013-9

Printed in the United States of America

10 9 8 7 6 5 4 3 2 1

Printed and bound by Hamilton Printing Company

To the memory of my Grandmoms

C-K.C.

For Dan and Ethel and Dan

G.E.K.

For my Parents

M.M., S.J.

Preface

Using this Guide

The purpose of this guide is to give a quick introduction on how to use *Maple*. It primarily covers *Maple* 7, *Maple* 8, and *Maple* 9, although we do note some references to previous versions of *Maple*, especially in cases where significant changes have been introduced. Also, throughout this guide, we will be suggesting tips and diagnosing common problems that users are likely to encounter. This should make the learning process smoother.

This guide is designed as a self-study tutorial to learn *Maple*. Our emphasis is on getting you quickly "up to speed." This guide can also be used as a supplement (or reference) for students taking a mathematics (or science) course that requires use of *Maple*, such as Calculus, Multivariable Calculus, Advanced Calculus, Linear Algebra, Discrete Mathematics, Modeling, or Statistics.

About *Maple*

Maple is computer algebra software developed by Waterloo Maple Inc. This software lets you use the computer like an interactive mathematics scratchpad. *Maple* can perform symbolic computation as well as numerical computation, graphics, programming and so on. It is a useful tool not only for an undergraduate mathematics or science major, but also for graduate students and researchers. The program is widely used as well by engineers, physicists, economists, transportation officials, and architects.

Organization of the Guide

The Guide is organized as follows:

- Chapter 1 gives a short demonstration of what you'll see in the remaining parts of the Guide. Chapters 2 through 10 contain the basic information that almost every user of *Maple* should know.
- Chapters 11 through 14 demonstrate *Maple*'s capabilities for single-variable calculus. This includes working with derivatives, integrals, series and differential equations.
- Chapters 15 through 20 cover topics of multivariable calculus. Here you'll find discussion of partial derivatives, multiple integrals, vectors, vector fields, and line and surface integrals.
- Chapters 21 and 22 introduce the statistical capabilities of *Maple*.
- Chapters 23 through 27 address a collection of topics ranging from animation and simulation to programming, list processing, and the use of spreadsheets.
- Five appendices explain how to ... work in the *Maple* worksheet environment ... use certain input features ... create your own *Maple* packages ... learn more about Calculus ... and work with scientific units and constants.

Chapter Structure

Each chapter of the Guide has been structured around an area of undergraduate mathematics. Each moves quickly to define relevant commands, address their syntax, and provide basic examples.

Every chapter ends with as many as three special sections that can be passed over during your first reading. However, these sections will provide valuable support when you start asking questions and looking for more detail. These three sections are:

- "More Examples." Here, you'll find more technical examples or items that address more mathematical points.
- "Useful Tips." This section contains some simple pointers that all *Maple* users eventually learn. We've drawn them from our experiences in teaching undergraduates how to use *Maple*.
- "Troubleshooting Q & A." Here we present a question and answer dialogue on common problems and error messages. You'll also be able to find out *how* certain commands work or *what they assume* you know in using them. Many useful topics are covered here.

Conventions Used in This Guide

Almost every *Maple* input and output you see in this guide appears exactly as we executed it in a *Maple* session. Slight changes were made only to enhance page layout or to guarantee better photo-offset production quality for graphics.

Also, our sections on "Useful Tips" are denoted with *light bulbs* ♉. More light bulbs indicate tips that we think are more important than others. Our scale is 1 to 4 light bulbs, but your wattage may vary.

Final Comments

We would like to thank Douglas B. Meade of University of South Carolina for giving us numerous suggestions and advice in our preparation of the second edition of the manual. His insight and knowledge of *Maple* have been a great help to us. Laurent Bernardin of Maplesoft also reviewed parts of this manuscript and offered useful advice, and we thank him for his comments.

Let us also thank all of our colleagues and students at Boston College who have contributed to the completion of this guide. Special thanks are extended to Jenny Baglivo, Nancy Gaff, and Sarah Quebec for their contributions to the statistics pages in the first edition of the guide. Bill Keane contributed several suggestions on the penultimate printing of the guide. Bill Zahner was extremely helpful for his reading, rereading, and critiquing of (many versions of) the guide as it developed. He also offered terrific suggestions on presentation and arrangement.

The staff at John Wiley, Inc., has been very supportive of our efforts, as well as patient with our schedule which always seemed to be slightly missing those deadlines we expected to meet. Our sincerest thanks go to Sharon Smith and Michael Boezi for their enthusiastic responses to this project.

Finally, despite our best efforts, it is likely that somewhere in these many pages, an error of either omission or commission awaits you. We sincerely apologize if this is the case and accept full responsibility for any inaccuracies.

We will provide a listing of further comments, examples, and any inaccuracies that may be found in this Guide. You can locate us and our websites by visiting our departments on the web at either:

> www.bc.edu/math   *or*   euler.slu.edu

We wish you only the best computing experiences in *Maple*.

C-K, Jerry, and Mike

Table of Contents

Part I. Basic *Maple* Commands

Part II. Drawing Pictures in *Maple*

Part III. *Maple* for One Variable Calculus

Part IV. *Maple* for Multivariable Calculus

Part V. *Maple* for Linear Algebra and Vector Calculus

Part VI. Using *Maple* in Statistics

Part VII. Advanced Features of *Maple*

Appendices and Index

CHAPTER 1

Running Maple

Computer Systems

What Computer System Are You Using?

Maple software runs on almost every major computer system including mainframes and desktop systems. *Maple* can also be set up to run across a network and even between systems.

Almost all implementations of *Maple* are graphically based systems where you can both type on a keyboard and use a mouse to navigate a window (e.g., desktop systems running MS Windows, UNIX-based systems with X-Windows software, Solaris, or the Mac system).

The *Maple* commands we discuss in this guide will work on all of these implementations.

> **Note:** Text-based implementations of *Maple* are available. These are often used on mainframe systems, and are usually of interest only to people who are doing remote logins, or to advanced users who want the performance advantage that comes from using a command line system.

Starting the Software

You should follow the instructions that came with the *Maple* software to install it on your computer system. Once you have completed the installation, you are ready to explore *Maple*.

Starting *Maple* obviously depends on the system you are using.

- On most PCs and Macintosh computers, you will typically find the icon of the *Maple* application in a window. Click (or double-click) on the icon.

- On UNIX-based, command-line systems, you will usually enter the command **xmaple** or **maple** (or a local equivalent, depending on how the software has been installed and how the system is configured).

  > **Note:** If you run *Maple* over a network, you may need to check with your system manager for the starting procedure.

The current version of *Maple* will be launched, and a new window will be opened with the cursor flashing, awaiting your input. At the top of the window, you will see the "menu bar," "tool bar," and "context bar". You will also see the *Maple* prompt, which is usually a greater-than sign ">". (Please see the graphic on the next page.) If you get an error message when starting *Maple*, check the Q & A section at the end of this chapter.

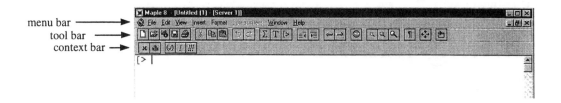

menu bar ⟶
tool bar ⟶
context bar ⟶

Input and Output

Maple is interactive software. For almost every entry you make, *Maple* will provide a direct response. Once you launch *Maple* and see the flashing cursor, you can type the four characters $\boxed{\textbf{1 + 1;}}$ and then press the evaluation key (either the **Enter** or **Return** key, depending on your computer system). *Maple* will give you the response "2" on a new line, and your window should look something like this:

```
[ > 1+1;
                      2
[ > |
```

> **Notes:**
>
> (1) *Maple* gave you the prompt symbol ">" automatically. You don't enter it yourself.
>
> (2) The semicolon is important!

(If you get an error message here, check the Q & A section.)

Now you enter the characters $\boxed{\textbf{39 - 11;}}$ and then press the evaluation key again (**Enter** or **Return**) and you will see the result of 28:

```
[ > 1+1;
                      2
[ > 39-11;
                     28
[ >
```

Throughout the rest of this manual, we will not show you windows or even the prompt sign >. Instead, we will show all of our input in boldface, with the response in plain text. Thus, our sequence above would be listed as:

1 + 1;
                2
39 - 11;
                28

Quit

If you have had enough and want to exit *Maple*, you can simply enter:

quit;

On graphically based systems, you can finish a *Maple* session by choosing the **Quit** item in the **File** menu.

A Quick Tour

For the rest of this chapter, we will show you some of *Maple*'s capabilities. We present these examples only to whet your appetite. You can follow along at your computer by typing (exactly!) what we show below. In later chapters, we will give you a more complete explanation of how to use these commands.

Note: When you input the following commands in *Maple*, make sure that:

- You use upper- and lower-case characters exactly as we do. *Maple* is "case sensitive." If you use the wrong capitalization, you can hurt *Maple*'s "feelings."

- You use exactly the type of brackets we show. There are three types of brackets: **[** *square brackets* **]**, **(** *parentheses* **)** and **{** *curly braces* **}**. Each has its own meaning in *Maple*. If you use the wrong one, *Maple* will be confused.

- You can continue input from one line to another in *Maple* by pressing either **Shift Return** or **Return** (depending on your system) as long as you do not enter a semicolon. For our examples, we recommend that you break the line exactly as we show in the text.

- You must always press the evaluation key (**Enter** or **Return**) to see the output.

Calculator

Maple does all the work of a hand-held electronic calculator. You can enter numerical expressions and *Maple* will do the arithmetic:

```
235.567*441.235/623.45;
```
$$166.7181092$$

```
sin(0.3);
```
$$.2955202067$$

But *Maple* can go much further. Try this factorial computation!

```
289!;
```

2079866075306145164348895732262527092227125189083652864966524223171\
405760295930638776430109826354519132675660433931363055910963871453\
772379754931444766652739192303201763588723618347593740385428725846\
122572271049818916876323493243976023302916666394540247449307010665\
731331903556896427962603583291932028351318887861286895384890867131\
005449989591695585446014805881310771610743578696897019623882572957\
311572603371048376255338230572538584588079078669943174850854858995\
580594914244562856410918607028520410684463210865874698240000000000\
0000000000000000000000000000000000000000000000000000000000000000

You cannot get this on your calculator!

Solving Equations

Maple can solve complicated equations and even systems of equations in many variables. For example, the equations $2x + 5y = 37$ and $x - 3y = 21$ have the simultaneous solution:

```
solve( { 2*x+5*y = 37, x-3*y = 21 }, { x, y } );
```
$$\left\{ y = \frac{-5}{11}, \quad x = \frac{216}{11} \right\}$$

Maple can also find solutions to equations numerically. For example, the equation $x = \cos(x)$ has a solution very close to $x = 0.75$. We can find it with:

```
fsolve( x=cos(x), x );
```
$$.7390851332$$

You can learn more about solving equations in Chapter 5.

Computer Algebra

Maple is very good at algebra. It can work with polynomials:

```
expand( (x-2)^2 * (x+5)^3 );
```
$$x^5 + 11x^4 + 19x^3 - 115x^2 - 200x + 500$$

```
factor( x^5+11*x^4+19*x^3-115*x^2-200*x+500 );
```
$$(x-2)^2(x+5)^3$$

Are you impressed? *Maple* also knows standard trigonometric identities such as $\sin^2(x) + \cos^2(x) = 1$:

```
simplify( sin(x)^2+cos(x)^2 );
```
$$1$$

That one was easy, but you probably forgot that $\sec^2(x) - \tan^2(x) = 1$:

```
simplify( sec(x)^2-tan(x)^2 );
```
$$1$$

You can learn more about doing algebra in *Maple* in Chapter 4.

Calculus

Maple even knows a lot about calculus! We can find the derivative of the function $f(x) = x/(1 + x^2)$ with:

```
diff( x/(1+x^2), x );
```
$$\frac{1}{1+x^2} - 2\frac{x^2}{(1+x^2)^2}$$

A complicated integral such as $\int \frac{1}{1+x^3}\, dx$ is handled rather easily.

```
int( 1/(1+x^3), x );
```
$$\frac{1}{3}\ln(1+x) - \frac{1}{6}\ln(x^2 - x + 1) + \frac{1}{3}\sqrt{3}\arctan\left(\frac{1}{3}(2x-1)\sqrt{3}\right)$$

Chapters 11 through 14 demonstrate many of the Calculus capabilities of *Maple*. Appendix D also contains information on *Maple* commands useful for learning Calculus.

Graphing in the Plane

Maple does everything a standard graphing calculator does and does it better. For example, to see the graph of the function $f(x) = x/(1 + x^2)$ over the interval $-4 \le x \le 4$, you can use:

```
plot( x/(1+x^2), x=-4..4 );
```

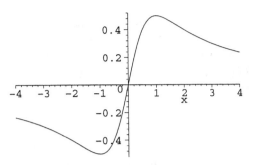

We can see a "daisy" with:

```
plot([cos(21*t)*cos(t), cos(21*t)*sin(t),t=0..2*Pi]);
```

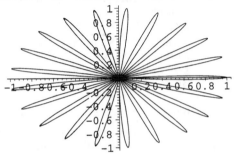

Chapters 8, 9, and 10 contain the details of the two-dimensional plotting capabilities of *Maple*.

Plotting in Space

Maple does a wonderful job with three-dimensional graphics. Let us show you two examples.

```
plot3d(sin(x)*cos(y), x=0..2*Pi, y=0..2*Pi);
```

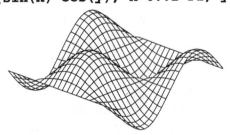

```
plot3d ([t/5, r*cos(t), r*sin(t)], r=0..1, t=0..6*Pi,
        grid=[8,60]);
```

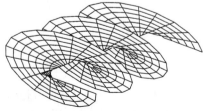

Chapters 15 and 19 contain most of the information you need to create three-dimensional pictures.

Changing the 3-D View

If you click and hold the (left) mouse button with the cursor/arrow at any point inside a 3-D picture, the picture will rotate according to how you move the mouse. This allows you to see the 3-D picture from any viewpoint.

Chapter 15 contains more information on how you can utilize *Maple*'s interface to more easily control your view of 3-D graphics.

Programming

Maple has its own programming language. You can use it to write code just as you would in BASIC, Pascal or C. Here is a simple program that uses the "do loop" to compute the sum of the squares from 1^2 to n^2, i.e., $1^2 + 2^2 + 3^3 + \ldots + n^2$:

```
squaresToN := proc(n)
      local sumThem, counter;
      sumThem := 0;
      for counter from 1 to n do
      sumThem := sumThem + counter^2;
      end do;
      return sumThem;
      end proc;
```

You can use this routine with statements such as:

```
print(`The sum of squares from 1-squared to 50-
squared is `, squaresToN(50));
```

The sum of squares from 1-squared to 50-squared is 42925

Chapter 27 introduces most of the basic features of the *Maple* programming language.

Simulation

You can use *Maple*'s programming language to do simulation. Here is a simple routine to simulate the flipping of a coin:

```
flip := proc()   local a;
a := rand(0..1)();
if a = 0 then "Head" else "Tail" end if;
end proc:
```

Here are eight coin flips:

```
seq(flip(), i =1..8);
```

"Head", "Tail", "Tail", "Tail", "Head", "Head", "Tail", "Head"

Chapter 25 contains more examples on random number and simulation.

More Examples

The "More Examples" sections of this guide present examples involving more mathematics. Students of mathematics, science, and engineering may find these of interest.

Here is one example. Not too many people know about the Bessel functions. But if you are learning physics, you might want to know what *Maple* has available for you in Bessel functions.

Special Functions

The Bessel function of order 0, $J_0(x)$, is a solution to the differential equation:

$$x^2 y'' + xy' + x^2 y = 0.$$

You can see its graph with:

`plot(BesselJ (0,x),x=0..20);`

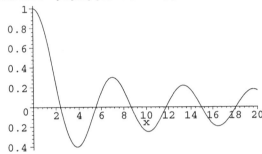

The smallest, positive zero of $J_0(x)$ is at approximately $x = 2.40483$:

`fsolve(BesselJ(0,x)=0,x);`

2.404825558

Useful Tips

 Maple is a memory-intensive application, so you should carefully compare the requirements of the software with the current amount of memory installed on your machine.

If the amount of memory on your system is tight, it is better to quit other applications before you start *Maple*. If you do need to run another program at the same time as you run *Maple*, start with *Maple* first.

Troubleshooting Q & A

Most chapters of this guide end with a Troubleshooting section, where we answer some common questions we think you will have. But there is not too much you can ask yet, except:

Question... When I tried to start *Maple*, I got an error message that there wasn't enough memory. What should I do?

Answer... *Maple* is memory intensive. Make sure that you have enough RAM installed on your machine to run the software. Also, it is a good practice to quit other applications before you launch *Maple*.

Question... If I forgot to end a *Maple* command with a semicolon, but hit **Return** or **Enter** anyway, what would happen?

Answer... Depending on your system, failing to add the semicolon can cause an error message, or the system may simply give a new prompt ">" and wait for you to finish the input.

Question... What do I do now?

Answer... Turn the page and start learning about *Maple*! Also, check Appendix A for how to use a *Maple* "worksheet".

CHAPTER 2

Calculator Features in Maple

Simple Arithmetic

Basic Arithmetic Operations

In this chapter, we will show you how *Maple* can work as a calculator. We begin with the basic arithmetic operations used in *Maple*:

Command	What It Does
+, −	Add, Subtract
*, /	Multiply, Divide
^	Raise to a power (exponentiation)

Here are some expressions to evaluate: $\dfrac{25.5}{5}$, $4 + 2^5$ and $\dfrac{23}{5} - \dfrac{3}{5} + 5(2^2)$:

```
25.5/5;
```
      5.100000000

```
4+2^5;
```
      36

```
23/5 - 3/5 + 5*2^3;
```
      44

We can also use parentheses to group terms together. For example, the expression $(3+4)\left(\dfrac{4-8}{5}\right)$ is entered with:

```
(3+4)*((4-8)/5);
```
$$-\frac{28}{5}$$

> **Note:** Use (*parentheses*) to group terms in expressions. Do not use [*square brackets*] or { *curly braces* } -- they mean something different.

Precedence

Maple follows the laws of precedence of multiplication over addition and so on, just as you do by hand. For example,

9/6*22+5;

> 38

actually computes $(\frac{9}{6} \times 22) + 5$. A common mistake is to think of the input as either $\frac{9}{6 \times 22} + 5$ or $\frac{9}{6} \times (22 + 5)$.

Comments

You can add a comment to any expression by starting the comment with the pound sign **#**. For example:

27*3;   # This multiplies 27 and 3.

> 81

Maple will ignore the phrase **This multiplies 27 and 3**. It is for your own reference.

We will use comment lines throughout this guide to write short reminders about what we are emphasizing in certain examples.

Previous Results and % Syntax

As a convenience, *Maple* lets you use the percentage sign, **%**, to stand for "the last result obtained." In this way, you can avoid retyping output when you want to work with it in your next input.

For example,

3200*12;

> 38400

% - 6500;   # Same as 38400 - 6500.

> 31900

(1000 + %)^2; # Same as (1000+31900)^2.

> 1082410000

You can use two percentage signs **%%** for the "second-last result obtained," and three percentage signs **%%%** for the third-last. However, this convention does not work with more than 3 percentage signs.

> **Notes.** Prior to *MapleV*, *Maple* used the double quotation character, **"**, instead of **%**, to mean "the last result obtained." We will make no further mention of this in the text.

Output Styles

Numeric Output

Maple gives you an exact (symbolic) value for almost every numeric expression:

(3+9) * (4-8) / 1247 * 67;

$$\frac{-3216}{1247}$$

You can force *Maple* to give you an answer that looks like the decimal answer you'd get on a calculator by using **evalf** (for floating point evaluation) with parentheses around an expression:

evalf((3+9) * (4-8) / 1247 * 67);

-2.578989575

You can see more digits in the answer – say 40 – with:

evalf[40]((3+9) * (4-8) / 1247 * 67);

$- 2.578989574979951884522854851643945469126$

Results like these are called **approximate numeric values** in *Maple*.

Scientific Notation

Maple uses a modified standard scientific notation to display results when the numbers either get very large or very small:

evalf(1234567890);

$.1234567890 \ 10^{10}$

0.000003492836 ;

$.3492836 \ 10^{-5}$

The spaces in the output above represent multiplication.

Built-in Constants and Functions

Built-in Constants

The mathematical constants used most often are already built into *Maple*. Be careful using upper-case and lower-case characters when you use these constants.

Constant	Value	Explanation	Maple
π	3.1415926...	Ratio of a circle's circumference to its diameter	**Pi**
e	2.71828...	Natural exponential	**exp(1)**
i	$i = \sqrt{-1}$	Imaginary number	**I**
∞	∞	(Positive) infinity	**infinity**

For example, to see the value of π to 45 significant digits, we use:

evalf[45](Pi);

3.14159265358979323846264338327950288419716940

To compute the numerical value of $\pi^4 - 5e^{1/3}$, we type:

evalf(Pi^4 - 5*exp(1/3));

90.43102896

Built-in Functions

Maple has many built-in functions. Here are the functions that you will probably use the most.

Function(s)	Sample(s)	Maple *Name(s)*
Natural logarithm	$\ln(x)$	`ln(x)`
Logarithm to base a	$\log_a x$	`log[a](x)`
Exponential	e^x	`exp(x)`
Absolute value	$\lvert x \rvert$	`abs(x)`
Square root	$\sqrt{x}$	`sqrt(x)`
Trigonometric	$\sin(x),\ \cos(x),\ \ldots$	`sin(x),cos(x),tan(x),` `cot(x),sec(x),csc(x)`
Inverse trigonometric	$\sin^{-1}(x),\ \cos^{-1}(x),\ \ldots$	`arcsin(x),arccos(x),` `arctan(x),arccot(x)`, etc.

Hyperbolic	$\sinh(x),\ \cosh(x),\ \ldots$	`sinh(x),cosh(x),`
		`tanh(x),coth(x),`
		`sech(x),csch(x)`

Inverse hyperbolic	$\sinh^{-1}(x),\ \cosh^{-1}(x),\ \ldots$	`arcsinh(x),arccosh(x),`
		`arctanh(x),arccoth(x),`
		etc.

For example,

```
sin(Pi);
```
$$0$$

```
evalf(sin(180));    # 180 radians, NOT 180 degrees.
```
$$-.8011526357$$

```
arctan(1);
```
$$\frac{1}{4}\pi$$

```
exp(ln(exp(1)));
```
$$\mathbf{e}$$

> **Notes:**
>
> (1) *Maple* uses radian measure for all angles.
>
> (2) *Maple* refers to e^1 as **e** in the output. Unfortunately, you cannot use **e** when you enter input. You have to write **exp(1)** for input. (This behavior is consistent with what you typically find on a graphing calculator where there is an e^x button, but not an e button.)

Error Messages

If you make a mistake in your input, *Maple* will *print an error message* and the cursor will move to the place where *Maple* thinks the mistake is located.

Common Mistakes

The most common mistakes for beginners usually involve mismatching (or omitting) parentheses and forgetting to write the multiplication symbol, as you see in the following examples:

```
> 3*(4-5))+6;  # too many right perentheses
Error, `)` unexpected

> sin 3;  # it should be sin(3);
Error, unexpected number

> 5+2x^2  # we forgot the * and ; in 5+2*x^2;
Error, missing operator or `;`
```

More Examples

Approximate Numbers and Exactness

Working with **approximate numeric values** is just like working with values on a hand-held calculator. These numbers sometimes lose precision as values get rounded off in arithmetic.

■ **Example**. *Maple* makes a big distinction between the exact number π and a numerical approximation for it. For example, **sin(517*Pi)** is an exact quantity with an exact answer:

sin(517*Pi);

$$0$$

If you use a numerical approximation for 517π, you do not get an exact zero:

sin(evalf(517*Pi));

$$-.9407689571 \ 10^{-7}$$

This is very close to zero (it is, after all, -0.00000009407689571) and it is probably acceptable for the work you will be doing. But it is not exact.

Useful Tips

👁 👁 👁 👁 Don't forget to type the asterisk (*****) when multiplying terms. This is the most common mistake that beginners make. For example, you might incorrectly enter **23cos(0)** instead of **23*cos(0)**.

👁 👁 👁 👁 Never try to multiply terms together in *Maple* using only **(** *parentheses* **)**. For example, in mathematical writing you can write (3)(4) to denote 3×4. However, *Maple* has a different interpretation.

> **(3)(4);**
>
> 3

👁 👁 👁 Use parentheses in expressions to clarify what you mean. This helps avoid mistakes. For example, you might think that **3^2*x** means 3^{2x}, but it doesn't! It is actually $(3^2)x$ because the square is done before the multiplication of **2** and **x**. To get 3^{2x}, you should write **3^(2*x)**.

👁 👁 👁 Many of the *Maple* built-in functions (e.g., **sin**, **cos**, **ln**) are understood to be defined for *complex* arguments. As a result, you may sometimes get an unexpected result that involves a complex number.

In *Maple* 7 and later version, you can ask *Maple* to make its calculations using real numbers by including the command **with(RealDomain);** before using these functions.

👁 👁 We recommend that you use the **%** symbol only for short sequences of one or two calculations that you do not expect to repeat later on. Use the **%** symbol sparingly.

👁 Avoid using **exp(1)** together with **^** to describe an exponential function. For example, the expression e^{2x} is better written as **exp(2*x)**, rather than **exp(1)^(2*x)**.

Troubleshooting Q & A

Question... When I tried to evaluate a built-in function, *Maple* gave an error message, "Error, unexpected number." What should I check?

Answer... Make sure that you included parentheses when using *Maple* functions. For example, a common mistake for beginners is to type:

> **sin 2;**
>
> Error, unexpected number

The correct input should be **sin(2);**.

Question... When I entered **sin(2)**, *Maple* just gave me the same thing back again. It didn't evaluate it. Why?

> **sin(2);**
>
> $\sin(2)$

Answer... *Maple* always gives you an exact answer. When you write sin(2) in mathematics, you don't try to simplify it. Neither does *Maple*. However, if you want a decimal approximation for sin(2), use **evalf**.

> **evalf(sin(2));**
>
> .9092974268

Question... I cannot get a numerical value of π from *Maple* using **evalf**. Why?

Answer... It is a common mistake for beginners to enter π as **pi** instead of **Pi**. *Maple* will consider **pi** as the Greek letter "π" rather than the geometrical constant "π". Therefore, **pi** does not have any numerical value.

> **pi;**
>
> π
>
> **evalf(pi);**
>
> π
>
> **evalf(Pi);**
>
> 3.141592654

Question... When I tried to evaluate a built-in function, *Maple* just returned the input unevaluated. I then used **evalf**, but still *Maple* did not give me a numerical value. What should I check?

Answer... Check your spelling. Most likely, you have misspelled the name of a built-in function or constant. For example, you may have used **cosine(Pi)** or **cos(pi)** instead of **cos(Pi)**.

Question... I got the wrong answer when I entered a complicated arithmetic expression. What should I check?

Answer... Always use (*parentheses*) to keep your expressions manageable and readable. Sometimes you can get tripped up by not knowing the exact order in which expressions are evaluated. Parentheses make the order of evaluation clear.

For example, $(2^3)^4 = 4096$ and $2^{(3^4)}$ are very different numbers:

> **(2^3)^4;**
>
> 4096
>
> **2^(3^4);**
>
> 2417851639229258349412352

But if you do not use parentheses, you might not know which one gets computed. (In this case *Maple* will not evaluate the ambiguous expression.)

```
2^3^4;          # The order of operations is not clear.
```
Error, `^` unexpected

Question... I got the error message "Error, missing operator or `;`" when I entered a complicated arithmetic expression. What should I check?

Answer... There are many possible errors that will generate this error message. You should:

- Check all the spelling.
- Check if all the (*parentheses*) are in the right places. In particular, make sure you correctly matched right and left parentheses.
- Check that you typed * for multiplication.

CHAPTER 3

Variables and Functions

Variables

Immediate Assignment

With a calculator, you can store a value into memory and recall it later. With a more advanced calculator, you can store several, different values under names such as A, B, C, and so on, and then use one or more of them in later calculations.

You can do even better in *Maple*. You can associate a name with any *Maple* expression or value and then recall it whenever you want. You do this using a colon followed by an equal sign **:=**, which is the symbol for **immediate assignment**. For example:

    a := 3.4;   # We assign **a** to have the value 3.4.

                $a := 3.4$

(Note that there is no space between the colon and equal sign.)

Once you've made an assignment, you can recall its value or use it in an expression:

    a;

        3.4

    a + 2;

        5.4

    a^2;

        11.56

How Expressions Are Evaluated

You may want to know how *Maple* keeps track of all the symbols and variables that you have defined.

Say we assign the name **myLunch** to the sum of **apple** and 3 times **banana**.

    myLunch := apple + 3 * banana;

        $myLunch := apple + 3\ banana$ # *Maple* spits back the definition
                                        # because **apple** and **banana**
                                          # have no associated values.

Now suppose we give the value 2 to **apple** and then reevaluate **myLunch**:

    apple := 2;   # Now **apple** has the value 2.

        $apple := 2$

  myLunch;

        $2 + 3\ banana$ # When *Maple* reevaluates **myLunch**,
                                # it substitutes the value 2 for **apple**.

If we now define the value of **banana** to be 3 and reevaluate **myLunch**:

banana := 3;   # Now **banana** has the value 3.

$banana := 3$

myLunch;

11   # When *Maple* reevaluates **myLunch**, it replaces
            # **apple** and **banana** by their respective values, and
            # simplifies the resulting expression. Bon appétit!!

Redefining and Clearing Symbols

Once a symbol name has been assigned a value or an expression, *Maple* retains the association until you redefine it, explicitly clear it, or end your session.

a := 3.4;   # We assign **a** to be 3.4.

$a := 3.4$

a := 5;   # We reassign **a** to be 5.

$a := 5$

a + 2;   # *Maple* uses the most recently assigned value of **a**.

7

The easiest way to ask *Maple* to forget about an assignment is use the **unassign** command.

unassign('a');   # Note that we use single quotes around **a**.
a;

a

You can also **unassign** assignments for several names at the same time, using one statement:

unassign('myLunch', 'apple', 'banana');

Rules for Names

Names you use can be made up of letters and numbers, subject to the following two rules:

- You cannot use a name that begins with a number. For example, **2app** is not an acceptable name because *Maple* will think that you forgot to type an operator between "**2**" and "**app**," and so will give you an error message.

- You cannot choose names that conflict with *Maple*'s own names. For example, you cannot name one of your own variables **sin**.

All of the following are examples of legitimate names that you could use:

a, m, p1, A, area, Perimeter, Batman, classsOf2002

> **Note:** *Maple* distinguishes upper case and lower case characters. For example, the names **Batman**, **batman**, and **batMan** are different.

Substitution and Evaluation Commands

You can substitute values into an expression without explicitly assigning values to the variables. The substitution command, **subs**, is used in the form:

subs(*list of substitutions using =* **,** *expression* **);**

For example, to substitute $x = 2$ and $y = 5$ into the expression $x^2 - 2xy$:

```
subs(x=2, y=5, x^2 - 2*x*y );
```

$$-16$$

You can also achieve this result by using the evaluation command, **eval**. It is used in the form:

> **eval(** *expression,* **{** *list of substitutions using = */ **});**

In the previous example, we want to evaluate the expression $x^2 - 2xy$ at $x = 2$ and $y = 5$:

```
eval(x^2 - 2*x*y, {x=2, y=5});
```

$$-16$$

One advantage of using the substitution (or evaluation) command is that the value you substitute (or evaluate) into a variable is temporary and is not remembered by *Maple*.

```
x;                # The value of x is unchanged by the previous substitution or evaluation.
```

$$x$$

> **Note:** If we only need to substitute one variable with a value, there is no need to use the curly braces, **{ }**, in the **eval** command. For example,
> **eval(x^2 - 2*x*y, x=2);**.

Finding the Larger Root of a Quadratic Equation

■ **Example**. We want to find the larger root of each of the following quadratic equations: $2x^2 + 5x - 6 = 0$ and $2x^2 + x - 3 = 0$.

The roots of a quadratic equation $ax^2 + bx + c = 0$, with $a \neq 0$, are found using the quadratic formula $(-b \pm \sqrt{b^2 - 4ac})/(2a)$. The larger root, when $a > 0$, is thus:

```
unassign('a', 'b', 'c', 'largerroot');
largerroot := (- b + sqrt(b^2 - 4*a*c) )/(2*a);
```

$$largerroot := \frac{1}{2}\frac{-b + \sqrt{b^2 - 4ac}}{a}$$

Here are the larger roots of each of the two equations:

```
subs(a=2, b=5, c=-6, largerroot);
```

$$-\frac{5}{4} + \frac{1}{4}\sqrt{73}$$

```
eval(largerroot, {a=2, b=1, c=-3});
```

$$-\frac{1}{4} + \frac{1}{4}\sqrt{25}$$

The above answer can actually be simplified to 1. (*Maple* does not notice that $\sqrt{25} = 5$.) You can tell *Maple* to do so by using the **simplify** command which will be discussed in detail in the next chapter.

```
simplify(%);
```

$$1$$

Functions

Defining Functions

Maple has many built-in functions such as **sqrt**, **sin**, and **tan**. You can define your own functions as well.

To define a function $f(x)$ in *Maple*, you use:

> **f := x -> *formula* ;**

The "arrow symbol" **->** is formed by entering the minus sign and greater-than sign together, with no space(s) in between.

For example, you can define the function $f(x) = x^2 + 5x$ in *Maple* with:

> **f := x -> x^2+5*x;**

$$f := x \rightarrow x^2 + 5x$$

Now, we can do some evaluations:

> **f(7.1);**

$$85.91$$

> **f(a);**

$$a^2 + 5a$$

> **f(x+1);**

$$(x+1)^2 + 5x + 5$$

> **f(f(y));**

$$(y^2 + 5y)^2 + 5y^2 + 25y$$

Functions with More Than One Variable

Functions may have more than one variable. A simple example is the computation of the average speed of an automobile.

If an automobile travels m miles in the span of t minutes, then its average speed in miles per hour is given by the expression $\frac{m}{(t/60)} = \frac{60m}{t}$. We then have a speed function " $f(m, t) = \frac{60m}{t}$ ":

> **avgSpeed := (m,t) -> 60*m / t ;**

$$avgSpeed := (m,t) \rightarrow 60\,\frac{m}{t}$$

If a distance of 45 miles is traveled in 30 minutes, the average speed will be 90 m.p.h.:

> **avgSpeed(45,30);**

$$90$$

More Examples

**Functions of
Split Definition**

Functions sometimes cannot be defined using a single formula. For example, the famous Heaviside function is defined by:

$$H(x) = \begin{cases} 1, & \text{if } x > 0 \\ 0, & \text{if } x \le 0 \end{cases}$$

To define this function in *Maple*, we can use the **piecewise** command as follows:

```
h := x -> piecewise( x > 0, 1, x <= 0, 0);
```

$$h := x \to piecewise(x > 0, 1, x \le 0, 0)$$

(The characters "**<=**" in the command mean ≤, less than or equal to.)

To use **piecewise** to define a function having two branches, write:

```
piecewise( condition1 , result1 , condition2 , result2 )
```

This means that if *condition1* is satisfied, then *result1* will be used; otherwise, *Maple* will use *result2* if *condition2* is satisfied. The following table shows some operators you will use to check conditions.

Operator	Meaning
=	Equal to
>	Greater than
>=	Greater than or equal to
and	And

Operator	Meaning
<>	Not equal to
<	Less than
<=	Less than or equal to
or	Or

The syntax for the **piecewise** command can also be extended. For example, the following function *f* has three branches and can be defined as you see below:

$$f(x) = \begin{cases} 1 - x, & \text{if } 1 < x < 3 \\ x^2, & \text{if } 0 \le x \le 1 \\ x + 2, & \text{if } x < 0 \text{ or } x \ge 3 \end{cases}$$

```
f := x -> piecewise( 1 < x and x < 3, 1-x,
       0 <= x and x <= 1, x^2, x < 0 or x >= 3, x+2);
```

To see this definition more clearly, we recommend that you write the conditions one on each line (using **Shift-Return** on some systems) and line them up carefully:

```
f := x -> piecewise( 1 < x and x < 3,     1-x,
                     0 <= x and x <= 1,  x^2,
                     x < 0 or x >= 3,    x+2);
```

Useful Tips

☿ ☿ ☿ ☿   Never assign values to the names **x**, **y**, **z**, or **t**. Otherwise, *Maple* will confuse them with the variables **x**, **y**, **z**, or **t** that you typically use when defining functions.

☿ ☿ ☿ ☿   You should always use **unassign** before defining functions. It will help you to avoid potential conflicts between variable names and function definitions.

☿ ☿ ☿   Most of *Maple*'s built-in names use only lower case letters. Several names introduced in later version begin with capital letters. If the names of your own variables and functions begin with a lower case letter and include a capital letter, (e.g. **myFunction**, **newVar**), you can easily distinguish them from *Maple*'s.

☿   You can define **e := exp(1);**. This will help to simplify many of your inputs. For example, $e^{0.3}$ can now be entered as **e^0.3**, instead of **exp(1)^0.3**. However, it will be less accurate than **exp(0.3)**.

Troubleshooting Q & A

Question... When I defined a new variable or function, I got the error message "attempting to assign to ... which is protected." What did I do wrong?

Answer... If you try to rename a built-in function or built-in constant, *Maple* will give you this error message. You cannot choose names that conflict with *Maple*'s own names.

Question... I tried to define a function, but I could not get it to work. What should I check?

Answer... Three things usually bother function definitions.

- First, check for the proper syntax. Make sure you use a colon-equal definition. Also, do not leave any space between the **–** and **>** keys in forming the **->** sign. (Common mistakes include using **=** instead of **:=** and using **f(x):=** instead of **f := x ->**.)

- Second, check that you have entered the formula of the function correctly. Some common mistakes are to use expressions such as **2x**, **cos x**, **e^x**, and so on.

- Third, your function or variable name may conflict with something else you used earlier in your *Maple* session. For example, you may have once defined:

      x := 3;

  Later you define:

      f := x -> x^2;
      f(x);

9

The result is not the x^2 you expected, but $3^2 = 9$ since **x** has the value 3. Make sure you explicitly clear the variable(s) before you define your function:

```
unassign('x');
f := x -> x^2;
```

Please note that even if you start a new worksheet in your current *Maple* session, all the variables or functions that you defined in earlier worksheets will still be retained until you quit *Maple*.

Question... I am still not sure about when I should use an equal sign **=** or when I should use a colon-equal **:=** definition?

Answer... The symbol **=** means equality, while **:=** means assignment. Equality is a test that gives a true or false answer. Assignment is an action that either gives a name to a value or defines a function.

Question... When I tried to evaluate a function, I got the error message "`... uses a 2nd argument, ... , which is missing.`" What went wrong?

Answer... This message means that you did not supply enough variables to evaluate the function.

A common mistake is to define a function of one variable, say,

```
f := x -> x^2;
```

Later, you use the same name **f** to define a function of two variables, say,

```
f := (x, y) -> x + 2*y;
```

The original definition of $f(x) = x^2$ has now been erased. If you type **f(x)**, you will get an error message complaining about not having enough variables (because *Maple* is now expecting two variables for the function).

Question... When I used **unassign**, I got the error message "`... cannot unassign.`" What went wrong?

Answer... Make sure that you use two single quotes around the variable name that you want to clear. For example:

```
a := 3;
```

```
unassign(a);          # This will give you an error message.
```
Error, (in unassign) cannot unassign `3'

```
unassign('a');        # This is correct.
```

CHAPTER 4
Computer Algebra

Working with Polynomials and Powers

In this chapter we will show you how common algebraic operations can be done directly in *Maple*. Let us start with polynomials.

The expand and factor Commands

The **expand** command does exactly what its name says it does:

```
expand((x-2)*(x-3)*(x+1)^2);
```

$$x^4 - 3x^3 - 3x^2 + 7x + 6$$

The **factor** command is basically the reverse of the **expand** command:

```
factor(x^4-3*x^3-3*x^2+7*x+6);
```

$$(x - 2)\ (x - 3)\ (x + 1)^2$$

Here are some more examples:

Example	Comment
`factor(x^2-3);` $$x^2 - 3$$	Although $x^2 - 3 = (x + \sqrt{3})(x - \sqrt{3})$, **factor** will not by default give radicals in its answer.
`factor(x^2-3,sqrt(3));` $$(x + \sqrt{3})\ (x - \sqrt{3})$$	We can give *Maple* a list of radicals that it is allowed to use in a factorization.
`expand((x-1.54)*(3.2*x-2.9));` $$3.2x^2 - 7.828x + 4.466$$ `factor(%);` $$3.2\ (x - .90625000)\ (x - 1.5400000)$$	**factor** does a nice job even when you use numerical coefficients.
`factor((5-3*I)+(-4+I)*x +` `   (1-I)*x^2);` $$\left(\frac{1}{2} - \frac{1}{2}I\right)(2x - 3 - 5I)\ (x - 1 + I)$$	**factor** works even when the coefficients are complex.
`factor(x^2+1);` $$x^2 + 1$$	**factor** won't use by default complex numbers unless at least one of the coefficients is a complex number.
`factor(x^2+1, I);` $$(x + I)\ (x - I)$$	You can force a factoring into complex terms by listing **I** as the optional second input.

`expand((x-y+z)^3);` $$x^3 - 3x^2y + 3x^2z + 3xy^2 - 6xyz +$$ $$3xz^2 - y^3 + 3y^2z - 3yz^2 + z^3$$ `factor(%);` $$(x-y+z)^3$$	The **expand** and **factor** commands can also work for polynomials with more than one variable.

The simplify Command

The **simplify** command tries to produce an expression that *Maple* thinks is the simplest form. Most times, this will be the same as what you think of as "simplest."

Consider, for example, $(x+1)^2 - 4x = x^2 - 2x + 1 = (x-1)^2$. The term on the left-hand side would not be considered simplified, and neither would the middle term. Let us see how the **simplify** command reacts to them.

    simplify((x+1)^2-4*x);

$$x^2 - 2x + 1$$

    simplify(x^2-2*x+1);

$$x^2 - 2x + 1$$

    simplify((x-1)^2);

$$(x-1)^2$$

In the first two cases, *Maple* thinks that $x^2 - 2x + 1$ is the simplest form already, but in the third case, *Maple* thinks that $(x-1)^2$ is the simplest form too!

The symbolic Option

By default, *Maple* is very careful in simplifying. It understands that `sqrt(x^2)` is not always the same as **x** for all complex numbers. Instead, `sqrt(x^2)` simplifies to **x** times the complex sign of **x**.

    simplify(sqrt(x^2));

$$\text{csgn}(x)\,x$$

(In the output, csgn stands for complex sign.)

Sometimes we want such expressions of powers simplified anyway. In those cases we use the **symbolic** option of the **simplify** command.

    simplify(sqrt(x^2), symbolic);

$$x$$

The **symbolic** option of the **simplify** command will handle $\sqrt{x^2}$ by treating it symbolically as $(x^2)^{1/2}$ and rewriting it to be $x^{2(1/2)} = x^1$.

Similarly:

    simplify((x^6)^(1/3));

$$\left(x^6\right)^{1/3}$$

    simplify((x^6)^(1/3), symbolic);

$$x^2$$

Working with Rational Functions

The simplify and convert Commands

A rational function is an expression of the form $\frac{a\ polynomial}{another\ polynomial}$. The following are three common algebraic operations involving rational functions.

- Combining terms over a common denominator can be done with the **simplify** command. For example, to combine $\frac{2}{3x+1} + \frac{5x}{x+2}$:

  ```
  simplify( 2/(3*x+1) + (5*x)/(x+2));
  ```

  $$\frac{7x + 4 + 15x^2}{(3x+1)\ (x+2)}$$

- Splitting up rational functions into partial fractions can be done with the **convert** command if you specify the **parfrac** option. For example, to split up $\frac{11x^2 - 17x}{(x-1)^2(2x+1)}$:

  ```
  convert((11*x^2-17*x)/((x-1)^2*(2*x+1)),parfrac,x);
  ```

  $$-\frac{2}{(x-1)^2} + \frac{3}{(x-1)} + \frac{5}{(2x+1)}$$

- The **convert** command also does long division when you use the **parfrac** option. For example, $(x^5 - 2x^2 + 6x + 1) \div (x^2 + x + 1)$ can be found with:

  ```
  convert((x^5-2*x^2+6*x+1)/(x^2+x+1),parfrac,x);
  ```

  $$x^3 - x^2 - 1 + \frac{2 + 7x}{x^2 + x + 1}$$

Working with Trig and Hyperbolic Functions

The **simplify**, **expand**, and **factor** commands can also be used for expressions that involve trigonometric functions, but sometimes their results may not come out the way you think they should.

Basic Trig and Hyperbolic Identities

The **simplify** command can recognize many basic trigonometric or hyperbolic identities:

```
simplify(sin(x)^2+cos(x)^2);
```

$$1$$

```
simplify(sin(x)^2-cos(x)^2);
```

$$-2\cos(x)^2 + 1$$

```
simplify(cosh(x)^2-sinh(x)^2);
```

$$1$$

As you can see in the second example above, *Maple* often "simplifies" trigonometric functions by using the cosine function instead of the sine function.

The **expand** command works on trigonometric and hyperbolic functions without any difficulties.

expand(sin(2*x));

$$2\sin(x)\cos(x)$$

expand(cosh(2*x));

$$2\cosh(x)^2 - 1$$

The combine Command

The **factor** command is not very effective in working with trig and hyperbolic functions. For example,

factor(sin(x)^2-cos(x)^2);

$$(\sin(x) - \cos(x))\ (\sin(x) + \cos(x))$$

Instead we need to use the **combine** command to produce better results.

combine(sin(x)^2-cos(x)^2);

$$-\cos(2x)$$

combine(2*sin(x)*cos(x));

$$\sin(2x)$$

combine(cosh(x)^2+sinh(x)^2);

$$\cosh(2x)$$

combine(sin(2*x)*cos(3*x));

$$\frac{1}{2}\sin(5x) - \frac{1}{2}\sin(x)$$

Useful Tips

 You have to be careful when you use the **symbolic** option in the **simplify** command. Most of the time you use it, you will be assuming that all the quantities you are working with are non-negative real numbers.

Troubleshooting Q & A

Question... When I used **simplify**, **expand**, or **factor** on a polynomial, I got a number. What went wrong?

Answer... First, it is possible that after expansion or simplification, all the terms in your polynomial canceled out and left you with a constant.

If this is not the case, check whether you assigned a value to the variable at some earlier time. For example, you may have assigned

x := 5;

earlier, then after awhile you typed:

```
expand((x-3)^2*(4*x+5));
```
$$100$$

You asked *Maple* to expand $(5-3)^2*(4*5+5) = 100!!$ To correct this, type:

```
unassign('x'):
```

and reenter the **expand** command.

Question... When I used **simplify** or **expand** on a simple polynomial, I got the wrong answer. What went wrong?

Answer... Check to see that you remembered to put an asterisk (*****) between terms being multiplied together. Parenthesized expressions do not get multiplied when they are written next to each other. Consider:

```
expand((x+1)(x+2)^2);
```
$$\mathbf{x}(x+2)^2 + 2\mathbf{x}(x+2) + 1$$

Maple interprets the expression **(x+1)(x+2)^2** as the function $(x+1)^2$ evaluated at $(x+2)$. This is not what you wanted to do.

Question... I tried **simplify**, **expand**, **factor**, **combine**, and several other commands on an expression, and I cannot get the type of expression I am looking for to come out. What should I do?

Answer... There are more advanced techniques for controlling the way *Maple* simplifies an expression. But the practical answer may be that the software just cannot get you to where you want to be using algebra alone. You may need to look for a different approach.

For example, *Maple* cannot simplify $\tanh^{-1}\left((e^x - e^{-x})/(e^x + e^{-x})\right) = x$. But by graphing $\tanh^{-1}\left((e^x - e^{-x})/(e^x + e^{-x})\right)$, you see that it looks like $y = x$. (Chapter 8 does graphing.)   You can also compute that $\tanh^{-1}\left((e^x - e^{-x})/(e^x + e^{-x})\right)$ has derivative 1, so that is almost enough to establish the identity. (Chapter 11 will show you how to do derivatives.)

CHAPTER 5

Working with Equations

Equations and Their Solutions

The solve Command for an Equation

Maple's **solve** command will solve an equation for an "unknown" variable. You use it in the form:

> **solve(** *an equation* **,** *variable to solve for* **);**

For example,

> **solve(2*x+5 = 9, x);**
>
> 2

Notice that equations in *Maple* are written using the equal sign "**=**".

You can check that $x = 2$ is the correct solution to the above equation, using substitution syntax:

> **subs(x=2, 2*x+5 = 9);**
>
> $9 = 9$

You can even have *Maple* check that this substitution gives the correct answer by evaluating the result as a Boolean expression

> **evalb(subs(x=2, 2*x+5 = 9));**
>
> *true*

This means that after the substitution $x = 2$, the left-hand side of the equation equals the right-hand side.

Here are a few more examples involving **solve**:

Equation	To Solve It in Maple	Comment
Solve $x^2 - 3x + 1 = 0$ for x	**solve(x^2-3*x+1 = 0, x);** $$\frac{3}{2} + \frac{1}{2}\sqrt{5}, \quad \frac{3}{2} - \frac{1}{2}\sqrt{5}$$ **evalf(%);** $$2.618033989, .381966011$$	Equations can have more than one solution. We can see a numerical answer with the **evalf** command.
Solve $x^3 + x^2 = -3x$ for x	**solve(x^3+x^2 = -3*x, x);** $$0, \ -\frac{1}{2} + \frac{1}{2}I\sqrt{11}, \ -\frac{1}{2} - \frac{1}{2}I\sqrt{11}$$	Here two of the solutions are complex. I stands for $\sqrt{-1}$.
Solve $y^2 - ay = 2a$ for y.	**solve(y^2 - a*y = 2*a, y);** $$\frac{1}{2}a + \frac{1}{2}\sqrt{a^2 + 8a}, \quad \frac{1}{2}a - \frac{1}{2}\sqrt{a^2 + 8a}$$	If the equation involves other variables, *Maple* will treat them as constants.

Solve $x^2 = xyt$ for x and y	`solve( x^2=x*y*t, {x, y});` $\{x = 0,\ y = y\},\ \{x = yt,\ y = y\}$	Use {curly braces}, if there is more than one variable to solve for.
Solve $x + \sin x = \cos x$ for x	`solve(x+sin(x) = cos(x), x);` (No output from *Maple*.)	No solution is returned when *Maple* cannot solve an equation.

> **Note:** The **solve** command works very well for equations involving polynomials. However, it doesn't have much success with trigonometric, exponential, logarithmic, or hyperbolic functions.

The solve Command for a System of Equations

The **solve** command can also be used to solve a system of equations. For two equations in two unknown variables, you use this form of the command:

> **solve({** *equation1, equation2* **}, {** *variable1, variable2* **});**

■ **Example.** To solve the equations $3x + 8y = 5$ and $5x + 2y = 7$ in the variables x and y, use:

> `solve({3*x+8*y = 5, 5*x+2*y = 7}, {x,y});`

$$\{x = \frac{23}{17},\ y = \frac{2}{17}\}$$

■ **Example.** To solve the equations $3xy - y^2 = -4$ and $2x + y = 3$:

> `solve({3*x*y - y^2 = -4, 2*x + y = 3}, {x,y});`

$$\{x = -\frac{1}{2}\text{RootOf}(5\_Z^2 - 9\_Z - 8, label = \_L1) + \frac{3}{2},$$
$$y = \text{RootOf}(5\_Z^2 - 9\_Z - 8, label = \_L1)\}$$

Maple gives the answer in terms of the roots of the polynomial $5z^2 - 9z - 8$. (The expression *label* $= \_L1$ indicates that the values of x and y are built from the same root of the polynomial in the answer.) Since the polynomial is quadratic, we expect two sets of solutions. To get a list of solutions, we use the **allvalues** command:

> `allvalues(%);`

$$\{\ x = \frac{21}{20} - \frac{1}{20}\sqrt{241},\ y = \frac{9}{10} + \frac{1}{10}\sqrt{241}\ \},$$
$$\{\ x = \frac{21}{20} + \frac{1}{20}\sqrt{241},\ y = \frac{9}{10} - \frac{1}{10}\sqrt{241}\ \}$$

■ **Example.** If you try

> `solve({x+y = 0, x+y = 1},{x,y});`

No solution is returned because there is no solution to this system of equations.

■ **Example.** We can also solve systems of equations involving more than two variables. For example:

> `allvalues(solve({x+2*y-z=1, x-y+z^2=2, y+x=2*z}, {x,y,z}));`

$$\{z = -2 + 2\sqrt{2},\ y = 3 - 2\sqrt{2},\ x = -7 + 6\sqrt{2}\},$$
$$\{z = -2 - 2\sqrt{2},\ y = 3 + 2\sqrt{2},\ x = -7 - 6\sqrt{2}\}$$

Numerical Solutions for Equations

Numerical Answers from solve

The **solve** command can also use efficient numerical techniques to approximate roots of polynomial and a few other simple functions. However, in order to do this, at least one of the coefficients in the equation has to be a floating point number. For example:

```
solve(x^5-x^3-1=0,x);
```

$$\text{RootOf}(\_Z^5 - \_Z^3 - 1, index = 1),\quad \text{RootOf}(\_Z^5 - \_Z^3 - 1, index = 2),$$
$$\text{RootOf}(\_Z^5 - \_Z^3 - 1, index = 3),\quad \text{RootOf}(\_Z^5 - \_Z^3 - 1, index = 4),$$
$$\text{RootOf}(\_Z^5 - \_Z^3 - 1, index = 5)$$

Now, if we replace "–1" with "–1.0" in the equation (i.e., replacing the exact value 1 with its numerical equivalent 1.0), then *Maple* will report numerical approximations for the answers:

```
solve(x^5-x^3-1.0 =0,x);        # use 1.0 instead of 1
```

$$-.9590477179 - .4283659563\,I,\quad -.9590477179 + .4283659563\,I,$$
$$.3407948662 - .7854231030\,I,\quad .3407948662 + .7854231030\,I,$$
$$1.236505703$$

This trick will also work for systems of equations that involve polynomials or simple functions.

```
solve({3*x*y - y^2 = 4, 2*x^2 + y = 9.0},{x,y});
```
                                    # use 9.0 instead of 9

$$\{x = -2.946194209, y = -8.360120631\},\quad \{x = 2.031216860, y = .7483161321\},$$
$$\{x = 1.614445936, y = 3.787128638\},\quad \{x = -2.199468588, y = -.6753241393\}$$

If you used the same **solve** command without decimal place in the equation:

```
solve({3*x*y - y^2 = 4, 2*x^2 + y = 9}, {x,y});
```

the solution would involve the roots of a cubic equation. Using **allvalues** to convert that answer would produce several pages of output!

Numerical answers from the fsolve Command

To find numerical solutions for equations in general, use the **fsolve** command. It has the same syntax as **solve**.

```
fsolve( an equation , variable to solve for );
```

and for systems of equations, use it in the form:

```
fsolve({ equation1, equation2}, { variable1, variable2 });
```

For example:

```
fsolve(x^5-x^3-1=0,x);
```
$$1.236505703$$

In general the **fsolve** command gives a single answer. However, if the equation involves only polynomials of one variable, you can indicate the number of roots to look for with the **maxsols** option. Specify the **complex** option if you want complex values returned as well.

```
fsolve(x^5-x^3-1=0,x, complex, maxsols=5);
```

$-.9590477179 - .4283659563\,I, \quad -.9590477179 + .4283659563\,I,$
$.3407948662 - .7854231030\,I, \quad .3407948662 + .7854231030\,I, \quad 1.236505703$

You can usually use **fsolve** to find solutions of a complicated equation whenever **solve** fails:

```
solve( exp(x^2) -50*x^2+3*x = 0, x);    # no output
solve( exp(x^2) -50*x^2+3*x = 0.0, x);  # no output
fsolve( exp(x^2) -50*x^2+3*x = 0, x);
```

$-.1154942111$

Guiding fsolve with a Range

fsolve with a Range in an Equation

Sometimes we are interested in a root other than the one that **fsolve** returns. We can guide **fsolve** by giving it a range in which to look for the solution. The form of the command becomes:

> **fsolve(** *equation , variable , range for the solution* **);**

■ **Example.** We are interested in using **fsolve** to find solutions to the trig equation $\tan(x) = x$, If we know that there is a solution in each of the intervals $-\pi/2 \le x \le \pi/2$, $\pi/2 \le x \le 3\pi/2$, and $3\pi/2 \le x \le 5\pi/2$, we can specify appropriate ranges. We then find three solutions.

```
fsolve(tan(x) -x = 0, x, -Pi/2..Pi/2);
fsolve(tan(x) -x = 0, x, Pi/2..3*Pi/2);
fsolve(tan(x) -x = 0, x, 3*Pi/2..5*Pi/2);
```

0
4.493409458
7.725251837

fsolve with a Range in a System of Equations

To solve a system of two equations for the two unknowns x and y, restricted to the intervals $x_0 \le x \le x_1$ and $y_0 \le y \le y_1$, you write:

> **fsolve({** *equation1, equation2* **}, {x,y},**
> **{x =** $x_0 .. x_1$**, y =** $y_0 .. y_1$**});**

■ **Example.** The system $y^2 - x^3 = 5$ and $y = x - 3\cos x + 4$ has a solution inside each of the intervals $-2 < x < -1$ and $1 < y < 2$. We can pinpoint it with:

```
fsolve({y^2-x^3=5,y=x-3*cos(x)+4},{x,y},
    {x=-2..-1,y=1..2});
```

$\{y = 1.769915245, x = -1.231437723\}$

More Examples

Extracting Solutions from the Results of solve

Sometimes you will want to work with the answers you get from **solve** and **fsolve** without having to retype them. This example will show you the two steps that will enable you to do so.

Consider the equation $x^2 - 3x + 1 = 0$. It has two solutions:

```
solve(x^2-3*x+1 = 0, x);
```

$$\frac{3}{2}+\frac{1}{2}\sqrt{5}, \ \ \frac{3}{2}-\frac{1}{2}\sqrt{5}$$

• Step 1. Make the output of the **solve** command the contents of a list and give the list a name. (We discuss lists in the next chapter.)

```
ans := [solve(x^2-3*x+1 = 0, x)];
```

$$ans := \left[\frac{3}{2}+\frac{1}{2}\sqrt{5}, \ \ \frac{3}{2}-\frac{1}{2}\sqrt{5}\right]$$

• Step 2. Identify each of the answers with **ans[1]** and **ans[2]**.

```
ans[1];
```

$$\frac{3}{2}+\frac{1}{2}\sqrt{5}$$

```
ans[2];
```

$$\frac{3}{2}-\frac{1}{2}\sqrt{5}$$

Let us check that **ans[2]** is really a solution to $x^2 - 3x + 1 = 0$:

```
simplify(subs(x=ans[2],x^2 - 3*x + 1));
```

$$0$$

How about:

```
(x - ans[1]) * (x - ans[2]);
```

$$\left(x-\frac{3}{2}-\frac{1}{2}\sqrt{5}\right)\left(x-\frac{3}{2}+\frac{1}{2}\sqrt{5}\right)$$

```
expand(%);
```

$$x^2 - 3x + 1 = 0$$

Useful Tips

fsolve works much more quickly using *numeric methods* than **solve** does using *algebraic methods*.

fsolve can deal with all kinds of equations, but it only gives a single answer in most cases. (It can be given a range in which to look for a solution, but that removes the automation.) **solve** will try to find up to 100 solutions, but in many cases, cannot solve the equations symbolically.

We think the best approach is to use **solve** and include a floating point number in one of the equations to find a numerical solution first. Then, work with the output!

☝ ☝   *Maple* can return all solutions to some transcendental equation when **_EnvAllSolutions** is set to true. For example:

solve(sin(x) = 1/2);   #Maple will give an answer in the first quadrant.

$$\frac{1}{6}\pi$$

_EnvAllSolutions := true:
solve(sin(x) = 1/2);   #Maple will give the general solution.

$$\frac{1}{6}\pi + \frac{2}{3}\pi\_B1\sim +2\pi\_Z1\sim$$

(In this output, $\_B1$ is a binary constant (0 or 1) and $\_Z1\sim$ is an integer constant.)

☝ ☝   The following table can help you remember the difference between various types of brackets. But beware – they are not interchangeable!

Syntax Element	*Purpose*	*Example*
(*parentheses*)	(i) Grouping terms in a computation	**(x^2+3)*(x-1)**
	(ii) Arguments of commands	**sin(x^2)**
[*square brackets*]	List or vector of objects	**[x, y, z]**
{ *curly braces* }	Set of objects	**{x-3 = y,** ** 5*x+y = 2}**
List [*square brackets*]	Indexing members of a list	**soln[1]**

Troubleshooting Q & A

Question... When I used **solve** or **fsolve**, I got an error message. What should I check first?

Answer... You should first check if you made a mistake in the input of the equation(s).

- Check that you entered the formula correctly. Common mistakes include misspelling names of the *Maple* built-in functions, and forgetting to type the multiplication symbol "*****".

- Did you remember to include the variable(s) to solve for in the input?

- If you are using more than one equation, make sure all the curly braces and commas are located in the right places.

Question... When I used **solve** or **fsolve**, I got the error message "a constant is invalid as a variable." What should I check?

Answer... Check that the variable or variables you are trying to solve for have no assigned values. For example, you may have earlier assigned:

```
x := 3 ;
```

Later, you try:

```
solve( x^2-1=0, x );
```

Error, (in solve) a constant is invalid as a variable, 3

You should have cleared the variable(s) before using **solve**.

```
unassign('x');
```

Question... I used **solve** but could not understand *Maple*'s output. It had the symbols %1, %2, and so on. What does this mean?

Answer... If the result from **solve** is too lengthy, *Maple* will use the symbols %1, %2, etc., to represent expansions of subexpressions that occur several times.

Question... I used **solve** and *Maple* gave me the output "RootOf" What does this mean?

Answer... For example, if you try:

```
sol := [solve( sqrt(x) = (x-3)^2 , x )];
```

$$sol := [\ \mathrm{RootOf}(-\_Z + \_Z^4 - 6\_Z^2 + 9, \ 2.110124849)^2,$$
$$\mathrm{RootOf}(-\_Z + \_Z^4 - 6\_Z^2 + 9, \ 1.354977808)^2\ ]$$

Maple tells you that the solutions to the equation $\sqrt{x} = (x-3)^2$ are the squares of the roots of the polynomial $-z + z^4 - 6z^2 + 9$ near $z = 2.11$ and 1.35. This is not helpful. However, you can see the numerical value of these answers with:

```
evalf({sol[1], sol[2]} );
```

$$\{4.452626878, 1.835964860\}$$

If the equation involves polynomial only, you can also use the **allvalues** command to convert the answer to a friendlier form:

```
allvalues(%);
```

Question... *Maple* did not give any output from **solve** or **fsolve**. Why not?

Answer... This usually means one of three things.

- *Maple* does not know how to solve your equation(s) with the **solve** command.
- If you used **fsolve** with a specified interval, *Maple* was unable to locate a root in that interval. Make sure that the interval(s) you gave contains a root.
- There is no solution to your equation or system of equations. A single equation might reduce to an absurdity (e.g., $x = x + 1$ reduces to $0 = 1$). A system of equations may be inconsistent (e.g., $x + y = 1$ and $x + y = 2$).

Question... When I used **fsolve**, I got an error message that says "...should use exactly all the indeterminates." or "...is in the equation, and is not solved for." What happened?

Answer... *Maple* cannot use a numerical method to approximate the solution, because your equation involves a constant that is not defined. For example, if **a** is not defined, *Maple* can do:

```
solve(x^2 = a, x);
```

But you will get an error message with:

```
fsolve(x^2 = a, x);
```

Question... **fsolve** failed to give me a solution to an equation in an interval where I know there is a solution. What happened?

Answer... **fsolve** uses a version of Newton's method to search for an answer. The procedure will not always find an answer. It is particularly vulnerable to functions whose graphs have narrow spikes on otherwise well-behaved regions.

For example, away from the origin, $\tan(x) = x^3$ has this feature, and it is hard for **fsolve** to locate a root without a very precise range.

Question... After solving an equation, I get real and complex roots. How can I only select the real roots?

Answer... You can use the command

```
remove( has, your list, I );
```

to remove from *your list* any term that includes the expression "**I**". As a result, any term with the imaginary number **I** will be removed. For example,

```
soln := [solve( x^7-5*x+4 = 0.0, x)];
```

> soln := [1., -1.409402287, -.7607643824-1.155768213*I, -.7607643824+1.155768213*I, .5229014897-1.183774264*I, .5229014897+1.183774264*I, .8851280726]

```
realSoln := remove( has, soln, I );
```

> realSoln := [1., -1.409402287, .8851280726]

Only the real solutions are selected.

Similarly, if you want to select those roots that have non-zero imaginary part, you can use:

```
imaginarySoln := select( has, soln, I );
```

CHAPTER 6
Sets, Lists, and Sequences

Lists and Sets

What Is a List?   A **list** in *Maple* is an expression in which elements are separated by commas and enclosed in [*square brackets*]. For example, each of the following is a list.

    [5,4,3,2,1];   # Each element is a number.

          [5, 4, 3, 2, 1]

    [1,2,2,3,3,3,4,4,4,4,5];   # Each element is a number.

          [1, 2, 2, 3, 3, 3, 4, 4, 4, 4, 5]

    [3, [1,2], 6, \`good\`, \`bad\`, -2, \`ugly\`];

                 # The elements consist of number, list of numbers and names.

          [3, [1, 2], 6, *good, bad*, -2, *ugly*]

What Is a Set?   A **set** in *Maple* is an expression in which elements are separated by commas and enclosed in { *curly braces* }. For example, each of the following is a set.

    {5,4,3,2,1};   # The ordering of the elements in a set doesn't matter.

          {1, 2, 3, 4, 5}

    {1,2,2,3,3,3,4,4,4,4,5};   # Duplicate elements are removed.

          {1, 2, 3, 4, 5}

    { 3, [1,2], 6, \`good\`, \`bad\`, -2, \`ugly\`};

                 # The elements consist of number, list of numbers and names.

          {-2, 3, 6, [1, 2], *good, ugly, bad*, [1,2] }

The Difference between Sets and Lists   A **list** is an ordered set. Repetition of elements is allowed. On the other hand, ordering and redundancy of elements in a **set** does not matter. As we see in the examples above, the set **{5,4,3,2,1}** is the same as **{1,2,3,4,5}**, also duplicate elements are automatically removed from the set **{1,2,2,3,3, 3,4,4,4,4,5}**, but not from the list **[1,2,2,3,3,3,4,4,4,4,5]**,

Lists and sets are important structures in *Maple*. Many of *Maple*'s inputs and outputs are expressed using lists or sets. For example, *Maple* gives two sets for the output when we solve a system of equations:

    solve({x+y=2, x^2+y = 2}, {x,y});

          $\{x = 1, y = 1\}$, $\{x = 0, y = 2\}$

We can then collect the solutions to form a list:

$$\texttt{ans := [solve(\{x+y=2, x\char94 2+y = 2\}, \{x,y\})];}$$
$$ans := [\{x = 1, y = 1\}, \{y = 2, x = 0\}]$$

Sequences

The seq Command

The **seq** command can be used to generate a sequence of elements that can be defined by a mathematical formula. For example, to make a sequence of the form n^2, for each integer $1 \le n \le 16$, we will type:

seq(n^2, n=1..16);

$$1, 4, 9, 16, 25, 36, 49, 64, 81, 100, 121, 144, 169, 196, 225, 256$$

In general, the **seq** command is used in the form:

seq(*expression involving n*, n = n_0..n_1);

Here are a few examples:

Maple *Command*	Remark
seq(n^2, n=-2..8); $\quad$ 4, 1, 0, 1, 4, 9, 16, 25, 36, 49, 64	The variable n runs from -2 to 8.
seq([cos(n),n/(n+1)], n = 1..3); $\left[\cos(1), \dfrac{1}{2}\right], \left[\cos(2), \dfrac{2}{3}\right], \left[\cos(3), \dfrac{3}{4}\right]$	The expression in the **seq** command can be a list itself. In this case, we form a table of coordinate pairs.
seq(x^k,k=0..7); $\quad 1, x, x^2, x^3, x^4, x^5, x^6, x^7$	Each element can be a symbolic expression.

By combining the **seq** command with [*square brackets*] or { *curly braces* }, we can create a list or set. For example,

[seq(n, n=1..22)];

$$[1, 2, 3, 4, 5, 6, 7, 8, 9, 10, 11, 12, 13, 14, 15, 16, 17, 18, 19, 20, 21, 22]$$

{seq(cos(n), n =1..5)};

$$\{\cos(5), \cos(4), \cos(3), \cos(1), \cos(2)\}$$

More Examples

An Experiment in Factoring

Let us look at the factorization of $x^n + 1$. With the help of the **seq** command, we can see the factorization for several values of n very quickly, say for $n = 2, 3, \ldots, 10$.

seq(factor (x^n+1), n=2..10);

$$x^2 + 1, \ (x + 1)(x^2 - x + 1), \ x^4 + 1, \ (x + 1)(x^4 - x^3 + x^2 - x + 1),$$

$$(x^2+1)\ (x^4-x^2+1),\quad (x+1)\ (1-x+x^2-x^3+x^4-x^5+x^6),\ x^8+1,$$
$$(x+1)\ (x^2-x+1)\ (x^6-x^3+1),\quad (x^2+1)\ (x^8-x^6+x^4-x^2+1)$$

Do you notice that all the coefficients in all the factors above are either 1 or –1? Try repeating the command with **n = 2..100**. You will see the same thing!

You may probably conclude that every coefficient in every factor of x^n+1 will be ± 1, no matter what n is. But before you celebrate your latest discovery, check the factorization of $x^{105}+1$. Surprise!

Drawing a Nonagon

■ **Example.** A nonagon is a regular polygon of nine sides (i.e., all nine sides have the same length and all nine angles are equal). It can be formed by joining together the points with coordinates $(\cos(2\pi n/9), \sin(2\pi n/9))$, where $n = 1, 2, \ldots, 9$.

We can create a list of these points using the **seq** command:

```
coordlist:=
        [seq([cos(n*2*Pi/9),sin(n*2*Pi/9)],n=0..9)];
```

$$coordlist := [[1,0],\ [\cos(\tfrac{2}{9}\pi),\ \sin(\tfrac{2}{9}\pi)],\ [\cos(\tfrac{4}{9}\pi),\ \sin(\tfrac{4}{9}\pi)],[\tfrac{-1}{2},\tfrac{1}{2}\sqrt{3}],$$
$$[-\cos(\tfrac{1}{9}\pi),\sin(\tfrac{1}{9}\pi)],\ [-\cos(\tfrac{1}{9}\pi),-\sin(\tfrac{1}{9}\pi)],\ [\tfrac{-1}{2},\tfrac{-1}{2}\sqrt{3}],$$
$$[\cos(\tfrac{4}{9}\pi),-\sin(\tfrac{4}{9}\pi)],\ [\cos(\tfrac{2}{9}\pi),-\sin(\tfrac{2}{9}\pi)],\ [1,0]]$$

If we join the points one by one, we will create the nonagon. We do this using the **plots[pointplot]** command. (**pointplot** will be discussed in Chapter 21.)

```
plots[pointplot]( coordlist, style=line,
                        scaling=constrained);
```

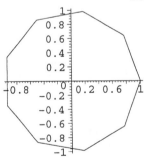

(The option **scaling=constrained** is used to make sure that both x- and y-axes have the same unit scale. We will explain this in detail in Chapter 8.)

Troubleshooting Q & A

Question... When I used the **seq** command, *Maple* gave me no output. What does that mean?

Answer... This means that the sequence you asked for is empty, because the index range you specified does not make sense. Make sure that $n_0 \le n_1$ in the command

```
seq( expression, n= n_0..n_1);
```

CHAPTER 7

Getting Help and Loading Packages

Text-based Help

Help for Specific *Maple* Commands

You can always get some help on how to use a command by typing:

> **? *command***

(In this case, it does not matter if you end the command with a semi-colon or not.) A help window will open, giving you details of the command, including options and examples.

For example, if you forget how to use the **fsolve** command, you can type:

?fsolve

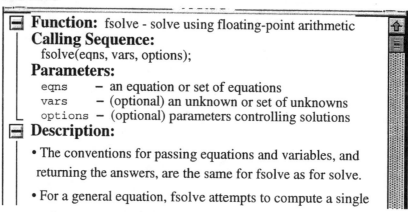

Function: fsolve - solve using floating-point arithmetic
Calling Sequence:
 fsolve(eqns, vars, options);
Parameters:
 eqns   – an equation or set of equations
 vars   – (optional) an unknown or set of unknowns
 options – (optional) parameters controlling solutions
Description:

• The conventions for passing equations and variables, and returning the answers, are the same for fsolve as for solve.

• For a general equation, fsolve attempts to compute a single

Looking Up a Command

Sometimes, you may not remember the exact name of a command that you want use. Say, you remember that it started with **plo**, but you just cannot recall its full name. You can ask *Maple*:

?plo

There are no matching topics. Try one of the following:
{plot, plot3d, plotdevice, plotoptions, plotouput, plots, plotsetup, plottools}

Browser-based Help

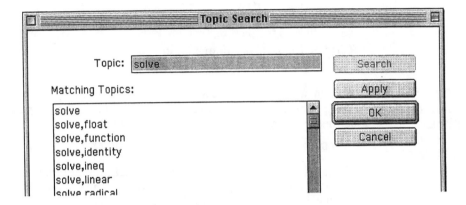

You can open up the interactive Help Window from the **Help** menu. To find information on a command or topic, you can either use the **Topic Search** or **Full Text Search** in the **Help** menu and select the appropriate topics. (The picture above shows what the Topic Search window looks on a Mac.)

The information displayed from the Help Window is the same as what you see when you use ?*command*.

> **Note:** Most of the help pages contain examples of how commands can be used. Pick the one or two sample commands that are closest to what you are trying to do, execute them, and see how they work.

Packages

Loading a Package

Maple has a built-in vocabulary of several hundred commands. But still these are not enough for everyday usage. Additional commands are available in the "library packages." These include commands for Algebra, Calculus, Linear Algebra, Number Theory, Statistics, and so on.

Before using those commands, you have to load their corresponding package using the **with** command. For example, the **animate** command is defined in the **plots** package. We will load the entire **plots** package with:

with(plots);

[animate, animate3d, animatecurve, arrow, complexplot, complexplot3d, conformal, conformal3d, contourplot, contourplot3d, coordplot3d, cylinderplot, densityplot, display, display3d, fieldplot, fieldplot3d, gradplot, gradplot3d, implicitplot, implicitplot3d, inequal, interactive, listcontplot, listcontplot3d, listdensityplot, listplot, listplot3d, loglogplot, logplot, matrixplot, odeplot, pareto, plotcompare, pointplot, pointplot3d, polarplot, polygonplot, polygonplot3d, polyhedra_supported, polyhedraplot, replot, rootlocus, semilogplot, setoptions, setoptions3d, spacecurve, sparsematrixplot, sphereplot, surfdata, textplot, textplot3d, tubeplot]

The output shows all of the commands defined in the **plots** package that have been loaded.

Now we can use the **animate** command:

```
animate(sin(2*x*t), x=-2..2, t=-2..2);
```

Now, click on the picture, then click the play button and the show starts! (Chapter 23 discusses animation in detail.)

Using a Command Without Loading a Package

You can also load and use a single command from a package without loading the whole package. For example, if you only want to use the **polygonplot** command defined in the **plots** package, you can type:

```
plots[polygonplot]( input values for the command );
```

Maple will then know where to find the definition of that command. However, since you have not loaded the **polygonplot** command, the next time when you use it, you have to type **plots[polygonplot]** again.

As another example, if you want to use the **mean** command which is defined in the **describe** subpackage of the **stats** package, you type:

```
stats[describe, mean]( input values for the command );
```

Useful Tips

When you load a package, finish the command with a colon instead of a semicolon,

```
with( package ):
```

Then *Maple* will not display the whole list of the commands in that package, which can be very lengthy.

CHAPTER 8

Making 2-D Pictures

Drawing the Graph of a Function

The plot Command

A common operation in mathematics is to plot the graph of a function, such as that of the square function, over a given interval. *Maple* does this with its **plot** command. To see the graph of $f(x) = x^2$ over the interval $-3 \le x \le 2$, type:

```
plot(x^2, x=-3..2);
```

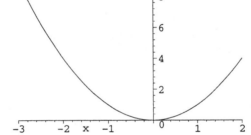

In general, to plot a function of x over an interval $a \le x \le b$, you type:

```
plot( function, x = a..b ) ;
```

For example, this plots the graph of $y = x^3 - 3x^2 + 5x - 10$ for x in the interval $-2 \le x \le 3$.

```
plot(x^3 -3*x^2 + 5*x -10,   x=-2..3);
```

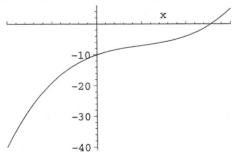

The smartplot command

An alternative method of plotting is with the **smartplot** command. You enter the function $f(x)$ needed to be graphed, then *Maple* will plot it over the interval $-10 \le x \le 10$.

```
smartplot( function );
```

43

For example, to see the graph of x^2, we type:

smartplot(x^2);

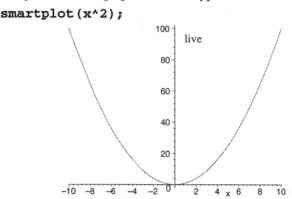

Notice the word "live" in the picture. This indicates that the **smartplot** picture is an interactive program. For example, you can select the output of a *Maple* computation and drag it to this **smartplot** picture. The graph of that expression will then be added to the picture.

In general we prefer to use the **plot** command over the **smartplot** command for graphing. The **plot** command will give us better control of the output graphics as you will see in the following.

Plot Options Context Bar

When you click once on the center of a picture from the **plot** or **smartplot** command, the context bar, near the top of the window, will display various plot options. By clicking various buttons in the options bar, you can select different plot styles, axes styles, and scales for the picture.

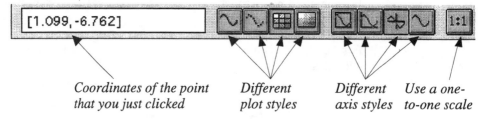

Coordinates of the point   Different   Different   Use a one-
that you just clicked   plot styles   axis styles   to-one scale

Plot Options Menus

When you click on the center of a picture, a number of extra menus will show up in the menu tool bar at the top of the worksheet, as you see in the picture below.

You can then ask *Maple* to redraw the graphic by choosing various options from these menus. The following table summarizes what each of the menus controls.

Menu	Comment
Style menu	Allows you to choose different styles for lines and points in a picture.
Legend menu	Lets you add and edit legends to a picture. This is most useful when the plot has more than one function.
Axes menu	Lets you choose different style for the axes. Also in the **smartplot** picture, there is a **range** option which allows you to adjust the ranges of the x and y variables.
Projection menu	Allows you to choose either **constrained** and **unconstrained** scalings. It serves the same purpose as the 1:1 button in the plot option tool bar.
Export menu	Lets you export the picture in various formats, such as postscript, gif, jpg, and so on.

> **Note**: You should use **constrained** scaling in the **Projection** menu or the 1:1 button on the plot option tool bar, whenever you want to see angles or circles properly. For example, a circle may look like an ellipse or the graph of $f(x) = x$ may not look like it makes a 45° degree angle with the x-axis unless you specify **constrained** scaling. (See the pictures below.)

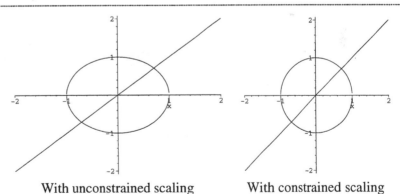

With unconstrained scaling   With constrained scaling

More Options with the Plot Command

Plotting Multiple Functions

The **plot** command lets you graph several functions or expressions simultaneously, all on the same set of axes, over a common interval $a \leq x \leq b$. Use the format:

```
plot ( { function1, function2, etc. }, x = a..b );
```

For example, to see the graphs $y = 3x^4 - 5x$, $y = 10\sin(x) - 10$, and $y = 5\cos(x) + 3e^x$, on the interval $-2 \leq x \leq 2$, you type:

```
plot({3*x^4-5*x, 10*sin(x)-10, 5*cos(x) + 3*exp(x)},
     x = -2..2);
```

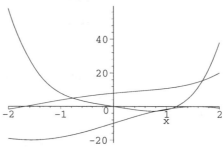

Notice that **{** *curly braces* **}** are used to group the functions together and that the functions share a same domain (in this case **x=-2..2**). Later in this chapter, we will show you another method to combine different pictures.

When drawing multiple functions, it may be better to use a legend to identify the graphs. You do this by choosing the **Show Legend** option in the **Legend** menu. You can also use the **Edit Legend** option to specify a legend.

Restricting the y-axis

If a graph contains a vertical asymptote or has a very large range of *y*-values, some of its interesting features may not be visible. That is because *Maple* tries to show you the whole graph. The solution is to restrict the portion of the *y*-axis you see.

For example, the graph of the tangent function has vertical asymptotes. Consider the difference between *Maple*'s default picture and one where we restrict the graph to show only *y*-values with $-10 \le y \le 10$.

```
plot( tan(x), x=-6..6 );
```

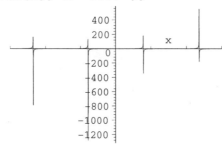

```
plot( tan(x), x=-6..6, y=-10..10 );
```

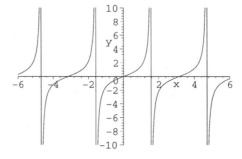

**Working with
Asymptotes
and
Discontinuities**

Maple draws asymptotes in the picture above because it actually tries to connect the branches of the graph together, from top to bottom. You can turn this behavior off by setting **discont = true** in the **plot** command.

```
plot( tan(x), x=-6..6 , y=-10..10, discont=true);
```

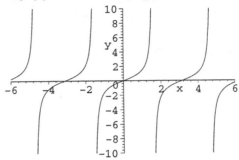

**Labeling
Pictures**

You can add a label to a picture and change the labels on the axes by using the **title** and **labels** options in the **plot** command.

```
plot( 15+cos(x), x=0..4*Pi ,
    labels = [ `day`, `price` ],
    title = `Daily price of Stock ABC` );
```

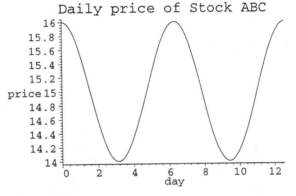

More Advanced Drawings

Colors

You can change the color of a picture by specifying the **color** option inside the **plot** command. You have a wide variety of colors to choose from, and *Maple* knows them under these names:

> **aquamarine, black, blue, navy, coral, cyan, brown, gold, green, gray, grey, khaki, magenta, maroon, orange, pink, plum, red, sienna, tan, turquoise, violet, wheat, white,** and **yellow.**

Say, let us draw a picture in orange:

```
plot(x^3+2*x, x=-3..2, color = orange);
```

Sorry, we cannot show you the picture here, because this text is printed in black and white!

Thickness

You can change the thickness of a graph by setting the **thickness** option to have the value 1 for a thin pen, 2 for a medium pen, or 3 for a thick pen. (The default is to use a thin pen.)

```
plot(x^3+2*x, x=-3..2, thickness = 3);
```

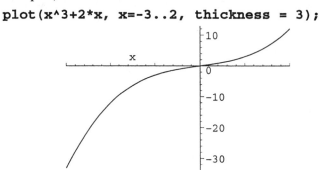

Combining Plots with the display Command

You can combine different graphics with different styles into a single picture by using the **display** command defined in the **plots** package.

First, name each of the graphics. Then, use the **display** command to combine these named graphics into a single output.

```
graph1 := plot( cos(x), x=-5..1, color=green):
graph2 := plot( sin(x), x=0..6,
                    thickness=2, color=yellow):
graph3 := plot( x^2-1, x=-1.5..1.5
                    thickness=3, color=red):
```

We also ended each command above with a colon instead of a semicolon. The colon serves the same purpose as a semicolon, except that the colon indicates that *Maple* should perform the desired calculation internally without displaying the result.

In this case, if we had used a semicolon, we could have still stored the plots, but then *Maple* would have displayed a long list of (unimportant) information from the **plot** commands.

```
with(plots):    # We have to use the plots package to use display.
display({graph1, graph2, graph3});      #or
display([graph1, graph2, graph3]);
```

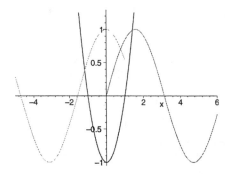

Notice that we end this command with a semicolon, because we want *Maple* to show us the picture. If you used a colon here, you will not see a picture.

Once you see the picture, you can then use the **Legend** Menu to show and edit legends.

More Examples

Zooming In

■ **Example.** Consider $f(x) = \sqrt{1 + 10x^4 - 20x^5 + 25x^6}$ and $g(x) = x^2 + 5\sin(x)$:

```
f := x -> sqrt( 1 + 10*x^4 - 20*x^5 + 25*x^6);
g := x -> x^2 + 5*sin(x);

plot({f(x),g(x)}, x=-3..3, y=-10..30);
```

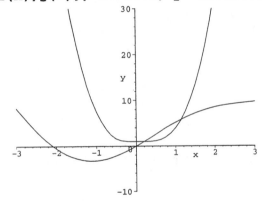

To get a better estimate of the intersection near 1.1, we can zoom in on the graph by successively shrinking down the *x*-interval in the **plot** command.

```
plot({f(x),g(x)}, x=1 .. 1.2);
plot({f(x),g(x)}, x=1.13 .. 1.15);
plot({f(x),g(x)}, x=1.135 .. 1.145);
```

An "Optical Illusion"

■ **Example.** Suppose we plot the function $f(x) = 64x^4 - 16x^3 + x^2$:

```
f := x -> 64*x^4-16*x^3+x^2;

plot(f(x), x=0..10);
```

You can easily get the impression that the graph is always increasing. However, notice that the vertical range is very large, between 0 and 600,000, so the picture is not sharp enough to show any small dips. In particular, the graph looks like a straight line for x between 0 and 2; this certainly is not the case.

```
plot(f(x), x=0..2);      # Still looks OK.
plot(f(x), x=0..1);      # Still looks OK.
plot(f(x), x=0..0.5);    # Still looks OK.
plot(f(x), x=0..0.2);    # Surprise!!
```

Useful Tips

☪ ☪ ☪ If you are trying to arrange several graphs with different plot options on the same graph, it is generally easier to draw the functions separately and then use the **display** command to put them together.

☪ ☪ ☪ *Maple* uses a variety of colors to draw 2-D graphs. These may look pleasing on the screen, but may not look very good when printed on a gray scale printer. We suggest that unless you have a color printer available, you add the option **color = black** to your plots (as we did when we printed the pages of this text).

☪ ☪ If you have a **plot** that involves various complicated functions, always try it first with one function at a time to make sure that each **plot** comes out OK by itself.

☪ ☪ You will find other methods of making 2-D pictures in Chapters 9 and 10. Also, there are many other options in **plot** that we have not discussed. You can find them using **?plot[options]** and experiment with them.

☪ ☪ If you use [*square brackets*] to group functions in the **plot** command, instead of { *curly braces* }, you can control the color and other presentation aspects of the graphs individually. For example, to draw the first graph in red and the second in blue, you can use:

```
plot( [x^2, x^3 ], x=-2..2, color = [red, green] );
```

☪ ☪ *Maple* 8 has an interactive **Plot Builder** that is a useful tool for learning graph options. To use the **Plot Builder**, the expression you want to graph must be a complete output expression. Click the output with the right mouse button (or hold down the option key when clicking for Mac users). A **Context** menu pops up. Select **Plot → Plot Builder**. A series of menu boxes will pop up. Select the options you want for your plot. *Maple* will automatically create the text command for producing the plot and show you the graph. The plot builder can also be invoked with the **interactive** command from the **plots** package. .

Troubleshooting Q & A

Question... I got an empty picture from **plot** with an error message about "*...empty plot.*" What went wrong?

Answer... This usually indicates that *Maple* cannot evaluate your input function numerically. Check whether you made a typo in the input. The common mistakes are:

- You typed the name of a built-in function incorrectly.
- You used the wrong variable.
- You specified an interval in which the input function is not defined.

Check whether your function really gives numbers! Do you get numbers when you enter **f(-1), f(0), f(1)** and so on?

Question... When I combined various pictures using the **display** command, I got a long list of numbers but no picture. Where is my mistake?

Answer... Make sure that you load the **plots** package before using the **display** command. Type:

```
with(plots);
```

and reenter your **display** command.

A second thing to check is that you used a colon rather than a semicolon when you defined the individual plots you are displaying together.

Also check that you have entered the format correctly. Note that you have to include all the names of the graphics inside a **{** *curly brace*s **}** or **[** *square brackets* **]** in the **display** command.

Question... When I combined various pictures using the **display** command, I got an error message. Where is my mistake?

Answer... First, make sure that you are not combining pictures from **smartplot**. (The **display** command will not work for **smartplot** picture.) Then, check if you typed the names of the pictures correctly in your **display** command. If there is no misspelling there, then you should **display** each picture one at a time in order to find out which one is causing the problem. Then recheck their definitions. (A common mistake is to use = instead of := in defining the pictures.)

Question... When I tried to combine various pictures from **smartplot** using **display**, I got an error message. Why is that?

Answer... You cannot combine the pictures from **smartplot** with **display**. Although the outputs from **smartplot** and **plot** appear to be similar, their structures are quite different.

The reason is that the pictures from **smartplot** allow several manipulations that are not offered in the standard plots. For example, you can control-drag an output expression into a **smartplot** picture then it will draw that expression. As a result, we cannot combine these pictures using **display** as we can with standard plots.

CHAPTER 9

Plotting Parametric Curves and Line Segments

Parametric Curves

Plotting Parametric Curves

In the previous chapter you saw how to use **plot** to draw curves that are graphs of functions. But not all curves are the graphs of functions.

A two-dimensional (2-D) parametric curve is written in the form $(x(t), y(t))$. The **plot** command can be used to draw the curve $(x(t), y(t))$, defined on an interval $a \leq t \leq b$. The command has the form:

```
plot([ x(t), y(t), t = a..b ]);
```

For example, to see the curve $(t^2 - 1, t + t^2)$ for $-2 \leq t \leq 2$, we use:

```
plot([t^2-1, t+t^2, t=-2..2]);
```

Plot Options

The options that we discussed in the last chapter (e.g., **scaling = constrained** and **color**) will also work for this particular format of **plot** command.

■ **Example.** To see the unit circle $(\cos t, \sin t)$ for $0 \leq t \leq 2\pi$, you use:

```
plot([cos(t),sin(t), t=0..2*Pi]);
```

Maple may give you a picture that does not look like a circle if the two axes are not measured in the same unit length. To change this, we add the option **scaling = constrained**. We will also draw it with a blue color:

```
plot([cos(t),sin(t), t=0..2*Pi],
        scaling = constrained, color = blue);
```

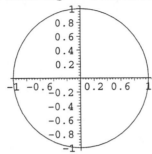

Plotting Line Segments in 2-D Plots

Line Segments   The **plot** command can also be used to plot a line segment joining the points (a, b) and (c, d). You type:

> plot([[a, b], [c, d]]);

or

> plot([[a, b], [c, d]], x = x_0..x_1, y = y_0..y_1);

For example, to see the line segment joining $(0, 1)$ to $(-1, 3)$:

```
plot([[0,1],[-1,3]], x=-2..2, y=0..3) ;
```

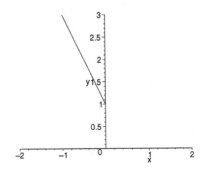

Plotting Multiple Graphics

Multiple Curves   You can plot several parametric curves, graphs, and lines all in one picture, using a single **plot** command. To do this, you put all the expressions of the curves inside { *curly braces* }.

■ **Example**. To draw the curves (t^2, t) for $-2 \le t \le 2$, and (t, t^2) for $-1 \le t \le 1$, the line segment from $(1, 0)$ to $(2, 3)$, and the graphs $y = 2x$, $y = x^2$ for $-2 \le x \le 3$, we can use:

```
plot({[t^2,t, t=-2..2] ,[t,t^2, t=-1..1],
     [[1,0],[2,3]], 2*x, x^2 }, x=-2..3);
```

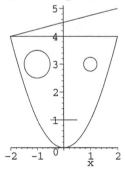

Multiple Pictures Together

You can also combine several 2-D pictures with the **display** command inside the **plots** package. The pictures you combine do not need to have the same interval for x. Also, we can assign different options to each individual picture.

■ **Example**. To create the "open skull" picture that you see below, we will combine several pictures. The face is made from a parabola and two straight lines; the eyes from two circles; and the mouth from a straight line.

```
face := plot({x^2, 4, 4.5+x/4},x=-2..2, color = red):
eyes := plot({[1+cos(t)/4, 3+sin(t)/4, t=0..2*Pi],
             [-1+cos(t)/2, 3+sin(t)/2, t=0..2*Pi]},
             color = brown):
mouth := plot([[-0.5,1],[0.5,1]], color = blue):

with(plots):
display({face,eyes,mouth}, scaling=constrained);
```

Useful Tips

♀ ♀ ♀   You may be overwhelmed by all the different plotting formats that we have discussed. The following table can help you to remember them:

Expression of the Form	What It Plots
`plot({ `$f_1(x), f_2(x)$` }, x = a..b);`	The graphs of $f_1(x)$, $f_2(x)$
`plot( [ `$x(t), y(t)$` , t = a..b] );`	The parametric curve $(x(t), y(t))$
`plot( [ [a, b], [c, d] ] );`	The line segment (a, b) to (c, d)

☼ ☼ If you need to draw a complicated picture that consists of various graphs or curves, always draw each picture individually and combine them with the **display** command. You may be able to use a single **plot** command to draw these graphics, but if you get an error message, it will be difficult to locate the mistake.

Troubleshooting Q & A

Question... I tried to draw a parametric curve with **plot** but got an error message. What went wrong?

Answer... This usually indicates that *Maple* cannot evaluate your input function numerically. Check whether you made a typo in the input. Common mistakes are:

- You typed the wrong name of a built-in function.
- You used the wrong variable.
- You specified an interval in which the input function is not defined.

Question... I tried to draw a parametric curve with **plot** but got an error message "`Error, (in plot) invalid arguments.`" What should I check?

Answer... This error message suggests that you should check the format of the **plot** command. Make sure that you used [*square brackets*] and (*parentheses*) correctly . For example, a common mistake is to input the expression `(t, t^2)` or `(t, t^2, t = -2..2 )` instead of `[t, t^2, t = -2..2]`.

Question... I tried to draw a parametric curve with **plot** but got an error message "... `parameter range must evaluate to a numeric.`" What should I check?

Answer... This error message suggests that you made a mistake in specifying the range for the parameter. For example, check to see if you used **pi** instead of **Pi**.

Question... I tried to draw one parametric curve with **plot**, but instead *Maple* gave me a picture of two curves. What went wrong?

Answer... When entering a parametric curve, it is common to forget to include the interval *inside* the [*square brackets*]. For example, instead of typing `plot([t, t^2, t = -2..2]);` you may make the mistake of typing:

```
plot([t, t^2], t=-2..2);
```

Then *Maple* will draw the *graphs* $y = t$ and $y = t^2$!! This is not what you want.

CHAPTER 10

Polarplot and Implicitplot

Plotting in Polar Coordinates

The polarplot Command

If a curve is expressed in polar coordinates, then we can draw it with the **polarplot** command. This command is defined in the **plots** package. (See the discussion of packages in Chapter 6.)

If the curve is given by $r = f(\theta)$, for $\theta_1 \le \theta \le \theta_2$, we can draw it with the following:

```
with(plots):
polarplot( f(θ), theta = θ₁ .. θ₂ );
```

Notice that this form is very similar to that of the **plot** command you already know. For example, to plot the three-leaf rose $r = 2\cos(3\theta)$, use:

```
with(plots):
polarplot( 2*cos(3*theta) , theta = 0..2*Pi);
```

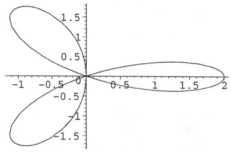

Plotting Multiple Curves with Style

Just as in the **plot** command, you can plot several curves by entering their equations inside { *curly braces* } or [*square brackets*]. Also, you can supply many of the options that work with the **plot** command.

For example, you can plot the spirals $r = \frac{\theta}{2\pi}$, $r = \left(\frac{\theta}{2\pi}\right)^2$ and the circle $r = 1$ all in one picture with:

```
polarplot({ theta/(2*Pi), (theta/(2*Pi))^2, 1 },
    theta=0..2*Pi, scaling=constrained, color=blue);
```

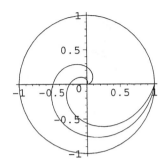

Plotting Graphs of Equations

The implicitplot Command

When a curve is given by an equation in the variables x and y, you can draw the curve with the **implicitplot** command (defined in the **plots** package).

You use the command in the form:

```
with(plots):
implicitplot( an equation in x and y , x = a..b, y = c..d );
```

For example, to see the unit circle $x^2 + y^2 = 1$ for $-1 \le x \le 1$, $-1 \le y \le 1$ use:

```
with(plots):
implicitplot( x^2 + y^2 = 1, x=-1..1, y=-1..1);
```

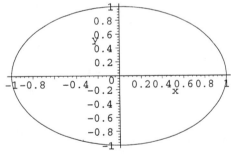

Multiple Curves and Styles

You can also use **{** *curly braces* **}** or the **display** command to plot multiple implicit curves. This is similar to the way you do it in the **plot** and **polarplot** commands. Many of the options you use with the **plot** command work with **implicitplot** as well.

■ **Example.** We want to see the curves $x^2 + 3xy + y^3 = 25$, $x^2 + 3xy + y^3 = 10$ and $x^2 + 3xy + y^3 = 0$ in the intervals $-10 \le x \le 10$ and $-6 \le y \le 4$. Let us draw each curve with a different color:

```
f := (x,y) -> x^2+3*x*y+y^3;
pict1 := implicitplot( f(x,y) = 25, x=-10..10,
         y=-6..4, color = red):


pict2 := implicitplot( f(x,y) = 10, x=-10..10,
         y=-6..4, color = yellow):
```

```
pict3 := implicitplot( f(x,y) = 0, x=-10..10,
                y=-6..4, color = green):
display({pict1,pict2,pict3});
```

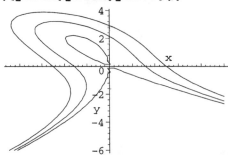

This picture looks like the Chinese character for "Wind." You can also produce this picture with the **contourplot** command that we will discuss in Chapter 16.

Useful Tips

☿ ☿   You may be overwhelmed by all the different plotting commands that we have discussed. Don't worry, because **plot** is the most commonly used command, while **polarplot** and **implicitplot** are for special situations:

To Draw	Use
Graph of a function $f(x)$	**plot**
Parametric curve $(x(t), y(t))$	**plot**
Curve in polar coordinates	**polarplot**
Curve given by an equation	**implicitplot**

☿ ☿   The **display** command lets you combine the results from different kinds of 2D-plot commands (such as **plot**, **polarplot**, **implicitplot**) on a single graph.

☿ ☿   Plots in polar coordinates can also be done with the **plot** command if you specify the **coords** option. Thus

```
polarplot( 2*cos(3*theta) , theta = 0..2*Pi);
```

and

```
plot([2*cos(3*theta), theta, theta=0..2*Pi],
                            coords=polar);
```

give the same result.

☿   *Maple* 7 (or higher) has an **algcurves** package that can be used for graphing polynomial equations implicitly. The **plot_real_curve** command in this package usually draws a more accurate curve, and with much more speed. For example:

```
implicitplot(x^2+y^2=1, x=-1..1, y=-1..1);
```

and

```
algcurves[plot_real_curve](x^2+y^2-1,x,y,
                      view=[-1..1,-1..1]);
```

draw the same curve.

Troubleshooting Q & A

Question... When I tried to use the **polarplot** or **implicitplot** command, *Maple* gave the command back without doing anything. What happened?

Answer... Usually, there are two reasons why this might happen:

- Check to see that you spelled the word **polarplot** or **implicitplot** correctly.
- You probably forgot to load the **plots** package before using these commands. Enter the following, and then try your commend again.

    with(plots);

Question... I got an error message "Plotting error, empty plot," when I used **polarplot**. What happened?

Answer... Check whether you made a typo in the input. Common mistakes are:

- You mistyped the name of a built-in function/variable.
- Your input function contains a variable other than **theta** (a common mistake is to include **r**, the radius variable, in the input).
- You did not match up parameters correctly (e.g., you used **t** in the function but used **theta** when you specified the interval).

Question... I got an error message from **implicitplot**. What should I look for?

Answer... There are two major problem areas in using **implicitplot**.

- Make sure that the equation you entered is really an equation with an equal sign "=". Check that the equation is entered correctly. Did you remember to type * for multiplication?
- Make sure that your equation has two variables that do not have values. Executing **unassgin('x','y');** before using **x** and **y** in your equation for **implicitplot** is highly recommended.

CHAPTER 11
Limits and Derivatives

Limits

The limit Command

If f is a function of a single variable, *Maple* evaluates $\lim\limits_{x \to a} f(x)$ with the following syntax:

```
limit( function, x = a );
```

For example, *Maple* agrees that $\lim\limits_{x \to 0} \dfrac{\sin x}{x} = 1$.

```
limit(sin(x)/x, x = 0);
```
$$1$$

The following table shows some sample limit computations, demonstrating that *Maple* can compute most limits, even those that involve infinite limits or limits at infinity:

Limit Calculation	*In* Maple
$\lim\limits_{x \to 0} \dfrac{e^x - 1 - x}{x^2} = \dfrac{1}{2}$	`limit((exp(x)-1-x)/x^2, x = 0);` $\dfrac{1}{2}$
$\lim\limits_{x \to 3} \dfrac{1-x}{(x-3)^2} = -\infty$	`limit((1-x)/(x-3)^2, x = 3);` $-\infty$

$\lim\limits_{x \to +\infty} \dfrac{x}{\sqrt{x^2 + 1}} = +1$	`f := x -> x/sqrt(x^2+1);`
	 `limit(f(x),x = infinity);`
	 1

$\lim\limits_{x \to -\infty} \dfrac{x}{\sqrt{x^2 + 1}} = -1$	`limit(f(x),x = -infinity);` -1
$\lim\limits_{x \to 0} \dfrac{\lvert x \rvert}{x}$ does not exist	`limit( abs(x)/x, x=0 );` *undefined*

One-sided Limits

To calculate a left-hand limit $\lim\limits_{x \to a^-} f(x)$, you have to add the **left** option in the **limit** command. To find $\lim\limits_{x \to 0^-} \dfrac{\sqrt{2x^2}}{x}$, you use:

```
limit( sqrt(2*x^2)/x, x = 0, left );
```
$$-\sqrt{2}$$

Similarly, to find the right-hand limit $\lim\limits_{x \to 0^+} \dfrac{\sqrt{2x^2}}{x}$, you use:

```
limit( sqrt(2*x^2)/x, x = 0, right );
```
$$\sqrt{2}$$

Differentiation

Differentiation Using the diff command

Maple uses `diff` for computing derivatives. You use it in the form:

```
diff( function, variable );
```

For example,

```
diff( x^4, x );
```
$$4x^3$$

To calculate a second derivative, you can use `diff` twice:

```
diff( diff( x^4, x ), x );
```
$$12x^2$$

Or you can use either of these short-hand formats:

```
diff( x^4, x, x );
```
$$12x^2$$

```
diff( x^4, x$2 );
```
$$12x^2$$

Similarly, the third derivative of the function f can be given by any of the following:

```
diff( x^4, x, x, x );      # Short-hand format.
```
$$24x$$

```
diff(x^4, x$3 );
```
$$24x$$

A few other examples of how you move from mathematical notation to *Maple* notation for derivatives are given in the following table.

Mathematical Expression	Maple *Evaluation*
$\dfrac{d(x^2 + e^{x^3})}{dx}$	`diff( x^2 + exp(x^3), x );` $2x + 3e^{x^3}x^2$
$\dfrac{d(\cos(y^2) + y^5)}{dy}$	`diff( cos(y^2) + y^5, y );` $-2\sin(y^2)y + 5y^4$

$$\left. \frac{df}{dt} \right|_{t=1}, \text{ where}$$

$$f(t) = t^2 + t^3 + \ln(t)$$

```
f := t -> t^2 + t^3 + ln(t):
eval(diff( f(t), t ), t=1 );
                6
```

In the last example above, we can also calculate $f'(1)$ quickly by using the **D** command. In *Maple*, when f is a function, the command **D(f)** means f', the derivative of f.

```
f := t -> t^2 + t^3 + ln(t);
```

```
D(f)(t);          # This calculates f'(t).
```

$$2t + 3t^2 + \frac{1}{t}$$

```
D(f)(1);          # This calculates f'(1).
```
$$6$$

Differentiation Rules

Maple knows all the formal computation rules in differentiation, such as:

```
unassign('f','g');        # Clears any previous definition of f and g.
diff(f(x)*g(x), x);       # The product rule
```

$$\left(\frac{\partial}{\partial x} f(x) \right) g(x) + f(x) \left(\frac{\partial}{\partial x} g(x) \right)$$

```
diff(f(x)/g(x), x);       # The quotient rule.
```

$$\frac{\frac{\partial}{\partial x} f(x)}{g(x)} - \frac{f(x)\left(\frac{\partial}{\partial x} g(x) \right)}{g(x)^2}$$

```
diff(f(x)^n, x);          # The power rule.
```

$$\frac{f(x)^n\, n \left(\frac{\partial}{\partial x} f(x) \right)}{f(x)}$$

How about the rule for differentiating the product of three functions?

```
diff( f(x)*g(x)*h(x), x);
```

$$\left(\frac{\partial}{\partial x} f(x) \right) g(x)\, h(x) + f(x) \left(\frac{\partial}{\partial x} g(x) \right) h(x) + f(x)\, g(x) \left(\frac{\partial}{\partial x} h(x) \right)$$

More Examples

Limits and Graphs

■ **Example.** We compute:

```
limit(sin(sin(2*x)^2)/x^2, x = 0);
```
$$4$$

We can check both numerically and graphically to see if this answer is correct.

- (*Numerically*) Calculate the values of the function at various test values of x that are close to zero. Check if they approach 4:

```
f := x -> sin(sin(2*x)^2)/x^2;

[f(0.12), f(-0.09), f(0.06), f(-0.025),
    f(0.01), f(-0.001), f(0.0001)];
```

$$[3.921699930, 3.956308721, 3.980700567, 3.996663621,$$
$$3.999466587, 3.999994668, 3.999999948]$$

- (*Graphically*) Check from the graph if the height of the function approaches 4, as x approaches zero.

```
plot( sin(sin(2*x)^2)/x^2, x = -0.5..0.5 );
```

| **Definition of Derivative** |

■ **Example.** Consider the function

```
f := x -> x^2*sin(x) + cos(x);
```

We know that the derivative is $f'(x) = \lim\limits_{h \to 0} (f(x+h) - f(x))/h$. When h is sufficiently small, say $h = 0.1$, we would expect $(f(x+0.1) - f(x))/0.1$ to be very close to $f'(x)$. We can see this by plotting these two functions on the same graph:

```
plot({diff(f(x),x), (f(x+0.1)-f(x))/0.1}, x=-3..3);
```

The result can be even better if we choose a smaller value for h, say $h = 0.01$:

```
plot({diff(f(x),x), (f(x+0.01)-f(x))/0.01},
    x=-3..3);
```

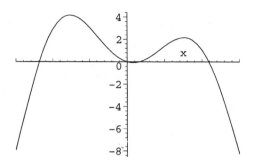

Geometry of the Derivative

■ **Example.** Consider

```
f := x -> 3*x^2 - 6*x*cos(x);
```

Then the value of the derivative at $x = a$, $f'(a)$, gives the slope of the tangent line at that point. Using this result, the equation of the tangent line is given by:

$$y = f(a) + f'(a)(x - a).$$

We can check this by combining the graph of $y = f(x)$ with the tangent lines at $x = -2.7$, $x = -1.4$ and $x = 0.9$. (Recall that **D(f)** means f'.)

```
graph1 := plot(f(x), x=-3.5..2, thickness=3):

line1 := plot(f(-2.7)+D(f)(-2.7)*(x+2.7),
            x=-3.5..-2):
line2 := plot(f(-1.4)+D(f)(-1.4)*(x+1.4), x=-2..0):
line3 := plot(f(0.90)+D(f)(0.90)*(x-0.90), x=0..2):

with(plots):
display({graph1, line1, line2, line3});
```

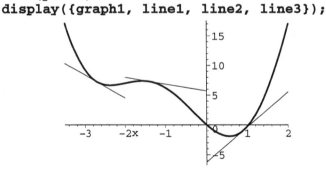

Useful Tips

  The commands **limit** and **diff** have inert versions that start with a capital letter (**Limit** and **Diff**). Each returns its argument unevaluated in mathematical notation.

If you are having trouble with a limit or derivative, start with the inert version of the command to make sure that you have the expression correct. Then change to lower case and evaluate the command. For example:

```
Diff( x*sin(x)+exp(x^2)+2*x, x) ;
```

$$\frac{\partial}{\partial x}\left(x\sin(x)+e^{(x^2)}+2x\right)$$

```
diff( x*sin(x)+exp(x^2)+2*x, x) ;
```

$$\sin(x)+x\cos(x)+2xe^{(x^2)}+2$$

```
Limit((exp(x)-1-x)/x^2, x=0)
        = limit((exp(x)-1-x)/x^2, x=0);
```

$$\lim_{x\to 0}\frac{e^x-1-x}{x^2}=\frac{1}{2}$$

Beginning with *Maple* 8, there is a **Student[Calculus1]** package that includes many commands to help students learn about Calculus. Please see Appendix D for more information on this package. (For users of earlier versions of *Maple*, the old **student** package included a number of commands that can help you if you are learning Calculus.)

Troubleshooting Q & A

Question... *Maple* returns my limit expression unevaluated. What does that mean?

Answer... This means that *Maple* cannot determine the value of the limit. You may be able to see what the limit is by looking at the graph of the function. (Recheck the first example in the More Examples section.)

Question... I keep getting zero for the derivative of a non constant function. What should I check?

Answer... You should check the following:

- Make sure that you are differentiating with respect to the correct variable. For example, you may have defined a function in terms of *t* but differentiated with respect to *x*.

- A common mistake is to type **diff(f, x);** to find the derivative $\frac{df}{dx}$. This is wrong, because this command means the differentiation of the *variable f* and not the expression $f(x)$. You have to type **diff(f(x), x);**.

Question... I cannot assign a new function which depends on the **diff** command. What should I look for?

Answer... The usual way of defining a new function using **:=** doesn't work very well with the **diff** command. For example, if you try

```
g := x -> diff(x^2, x) ;
g(x);
```

$$2\,x$$

It looks OK, but when you ask, say, for the value of g at $x = 2$

```
g(2);
```

Error, (in g) wrong number (or type) of parameters in function diff

This is wrong because *Maple* is trying to calculate **diff(2^2, 2)** which does not make sense. The correct way is to define g using the **D(f)** command.

```
f := x -> x^2;
g := x -> D(f)(x);
```

$$g := x \rightarrow 2x$$

```
g(2);
```

$$4$$

Another method is to use the **unapply** command:

```
unassign('g');        # Clear the previous definition of g.
```

$$g := g$$

```
g := unapply( diff(x^2, x), x);
```

$$g := x \rightarrow 2x$$

```
g(2);
```

$$4$$

CHAPTER 12
Integration

Antidifferentiation

The Integrate Command

You can use the **int** command to compute an indefinite integral $\int f(x)\,dx$. It has the form:

```
int( f(x), x );
```

You specify a function or an expression to integrate, as well as the variable in which the integration is to take place.

For example, to compute $\int x^2\,dx$, use this syntax:

```
int(x^2, x);
```

$$\frac{1}{3}x^3$$

> **Note:** *Maple* does not put the "+ C" in the answer of integration.

Maple can integrate almost every integral that can be done using standard integration methods (e.g., substitution, integration by parts, partial fractions). Here are some typical integrations:

Integral	In Maple	Comment
$\int x \ln(x)\,dx$	`int(x*ln(x), x);` $\frac{1}{2}x^2\ln(x) - \frac{1}{4}x^2$	This one uses integration by parts!
$\int \dfrac{y^2}{\sqrt{1-y^2}}\,dy$	`int(y^2/sqrt(1-y^2), y);` $-\frac{1}{2}y\sqrt{1-y^2} + \frac{1}{2}\sin^{-1}(y)$	This is computed using a trigonometric substitution (write $y = \sin u$).
$\int \sin(\cos(y^2))\,dy$	`int(sin(cos(y^2)), y);` $\int \sin(\cos(y^2))\,dy$	This integrand has no closed-form antiderivative.

When *Maple* cannot handle an integral, it usually returns your input unevaluated. This can mean either that it is not possible to find an antiderivative in closed form or that *Maple* has not yet been programmed to do the integral.

Definite Integrals

A definite integral $\int_a^b f(x)\,dx$ is computed in *Maple* with the **int** command:

```
int( f(x) , x = a..b );
```

Maple will try to find an antiderivative first, then evaluate it at the endpoints and subtract (according to the Fundamental Theorem of Calculus). Here are some examples:

Integral	*In* Maple	Comment
$\int_1^2 x^2 \, dx$	`int(x^2, x=1..2);` $\dfrac{7}{3}$	$\left.\dfrac{x^3}{3}\right\|_1^2 = \dfrac{8}{3} - \dfrac{1}{3} = \dfrac{7}{3}$
$\int_2^\infty \dfrac{1}{5+t^2} \, dt$	`int(1/(5+t^2), t=2..infinity);` $\dfrac{1}{10}\sqrt{5}\pi - \dfrac{1}{5}\sqrt{5}\tan^{-1}\left(\dfrac{2}{5}\sqrt{5}\right)$	This is an improper integral that involves $+\infty$.
$\int_0^1 \sqrt{\cos(x^2)} \, dx$	`int(sqrt(cos(x^2)), x=0..1);` $\displaystyle\int_0^1 \sqrt{\cos(x^2)} \, dx$	There is no antiderivative. *Maple* cannot evaluate it at the endpoints.

Numerical Integration

The evalf and Int Commands

We can use the **evalf** and **Int** commands together to find a numerical approximation for the integral $\int_a^b f(x) \, dx$ with the following syntax:

```
evalf( Int( f(x) , x = a..b ));
```

Note that **Int** by itself is the inert command for integration. This combination (**evalf** and **Int**) can calculate almost all definite integrals, including $\int_0^1 \sqrt{\cos(x^2)} \, dx$ for which **int** failed (as you saw earlier). Also, the computation is quick.

■ **Example.** To get approximate, numerical values for the integrals $\int_0^2 x^2 \, dx$ and $\int_0^1 \sqrt{\cos(x^2)} \, dx$, use these commands:

```
evalf(Int(x^2, x=0..2));
```
$$2.666666667$$

```
evalf(Int(sqrt(cos(x^2)), x=0..1));
```
$$.9485216211$$

More Examples

Area Between Curves

■ **Example.** To approximate the area bounded by the curves $p(x) = x^5 - 20x^3$ and $q(x) = 30 - x^5$, we start by sketching the curves. First, let us see where they intersect:

```
p := x -> x^5 - 20*x^3;
q := x -> 30 - x^5;

fsolve( p(x) = q(x), x);
```
$$-3.080038835, \quad -1.206322273, \quad 3.231778500$$

The values $-3.08, -1.206$ and 3.232 give approximations for the x-coordinates of the intersections. We can see the area between these two curves in the interval $-3.1 \le x \le 3.3$:

```
plot({p(x),q(x)}, x=-3.1..3.3);
```

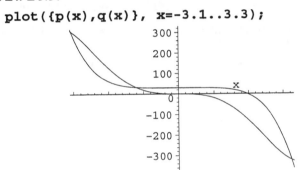

The area can be approximated with $\int_{-3.08}^{3.232} |p(x) - q(x)| \, dx$ using **evalf** and **Int**:

```
evalf( Int( abs( p(x)-q(x)), x=-3.08..3.232));
```
$$388.8535522$$

Fundamental Theorem of Calculus

■ **Example.** The Fundamental Theorem of Calculus states that if f is a continuous function, then $\dfrac{d}{dx} \int f(x)\,dx = f(x)$. So if you integrate a function f and then differentiate the result, you should expect to get f back again. Let us try it:

```
int( 1/(x^3+1), x);
```
$$\frac{1}{3}\ln(1+x) - \frac{1}{6}\ln(x^2 - x + 1) + \frac{1}{3}\sqrt{3}\tan^{-1}\left(\frac{1}{3}\sqrt{3}(2x-1)\right)$$

```
diff(%, x);
```
$$\frac{1}{3(1+x)} - \frac{1}{6}\frac{(2x-1)}{(x^2 - x + 1)} + \frac{2}{3 + (2x-1)^2}$$

It looks like *Maple* messed up! But wait . . .

```
simplify(%);
```

$$\frac{1}{(x+1)(x^2-x+1)}$$

which is of course the same as $\frac{1}{(x^3+1)}$.

Plotting an Antiderivative

■ **Example.** *Maple* cannot find an explicit formula for the antiderivative

$$\int 100\ln(x)e^{(-x^2)}\,dx.$$

```
f := x->100*ln(x)*exp(-x^2):
int(f(x),x);
```

$$\int 100\ln(x)e^{(-x^2)}\,dx$$

However, you can still plot its graph with the help of the **evalf** and **Int** commands. According to the Fundamental Theorem of Calculus, the function $F(x)$ = $\int_1^x 100\ln(t)e^{(-t^2)}\,dt$ is an antiderivative of $100\ln(x)e^{(-x^2)}$. *Maple* can compute values of F using numerical integration (**evalf** and **Int**) and you can then **plot** these values.

```
plot( evalf( Int(f(t), t=1..x) ), x=1..3 );
```

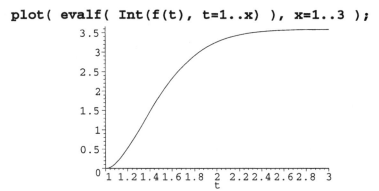

Useful Tips

💡 💡 Always use **evalf** and **Int** to evaluate a definite integral, unless you need an exact answer. In many cases, **int** will neither work nor give you a useful result. Even when **int** does work, it can be slow.

💡 💡 You can use the inert command **Int** to see whether you entered an integral correctly. (We mentioned the corresponding, inert commands **Limit** and **Diff** in the previous chapter.)

You can also use **Int** to form an expression with an integral in it, such as the following integration formula you'd see in a table of integrals:

```
Int(y^2/sqrt(1-y^2),y) = int(y^2/sqrt(1-y^2),y);
```

$$\int \frac{y^2}{\sqrt{1-y^2}}\, dy = -\frac{1}{2}y\sqrt{1-y^2} + \frac{1}{2}\arcsin(y)$$

☿　In Appendix D, we describe the **Student[Calculus1]** package in *Maple* 8 which includes many commands to help students understand integration.

☿　For users of earlier versions of *Maple*, the **student** package contains commands such as **Doubleint**, **intparts**, **leftsum** and **rightsum** that can help you if you are learning Calculus.

Troubleshooting Q & A

Question... I got some strange answers for definite integrals that used names like "**erf**," "**fresnelS**," and "**fresnelC**." What happened?

Answer... *Maple* knows about a lot of **special functions** that appear quite often in integration problems. For example, **erf** is the *Maple* name for the error function

$$erf(x) = \frac{2}{\sqrt{\pi}}\int_0^x e^{-t^2}\, dt$$

This is an integral that does not have a closed-form antiderivative, but its values are well known.

You might see something like this:

```
int(exp(-t^2), t=0..1);
```
$$\frac{1}{2}\sqrt{\pi}\; erf(1)$$

```
evalf(%);
```
　　　.7468241330

Maple has reported $\int_0^1 e^{-t^2}\, dt = \frac{\sqrt{\pi}}{2}(\frac{2}{\sqrt{\pi}}\int_0^1 e^{-t^2}\, dt) = (\frac{\sqrt{\pi}}{2})erf(1) \approx 0.747$.

CHAPTER 13

Series and Taylor Series

Series

The add Command

You can use the **add** command to add up a finite number of terms of an indexed expression. To find $\sum_{n=n_0}^{n_1} expression$, you type:

```
add( expression , n = n0..n1 );
```

For example,

```
add(n^2, n=1..20);          # Computes
```
$\sum_{n=1}^{20} n^2$.

$$2870$$

```
add(sin(n)/n, n=1..5);
```

$$\sin(1) + \frac{\sin(2)}{2} + \frac{\sin(3)}{3} + \frac{\sin(4)}{4} + \frac{\sin(5)}{5}$$

```
evalf(%);
```

$$.9621742221$$

The sum Command

However, if the summation involves symbolic computation, it is better to use the **sum** command instead. It has the same format as the **add** command:

```
sum( expression , n = n0..n1 );
```

For example, $\sum_{n=0}^{k} r^n = (1 - r^{k+1})/(1 - r)$ is a partial sum for a geometric series:

```
sum(r^n, n=0..k);           #use sum for symbolic series
```

$$\frac{r^{k+1}}{r-1} - \frac{1}{r-1}$$

```
add(r^n, n=0..k);           # add may fail for symbolic series
```

$$\text{Error, unable to execute add}$$

Infinite Series

The **sum** command can also be used to find an approximate value of certain infinite series. For example, $\sum_{n=1}^{\infty} \frac{1}{n^2}$ is known to converge:

```
sum( 1/n^2, n=1..infinity );
```

$$\frac{\pi^2}{6}$$

On the other hand, the harmonic series $\sum_{n=1}^{\infty} \frac{1}{n}$ diverges:

```
sum( 1/n, n=1..infinity );
```

$$\infty$$

Some Famous Infinite Series

■ **Example.** Here are some well-known infinite series:

$$\sum_{n=0}^{\infty} \frac{1}{n!} = e, \qquad \sum_{n=0}^{\infty} \frac{1}{(2n+1)^2} = \frac{\pi^2}{8}, \quad \text{and} \quad \sum_{k=1}^{\infty} \frac{k}{e^{2\pi k} - 1} = \frac{1}{24} - \frac{1}{8\pi}$$

Maple recognizes the first two symbolically but can only compute the third numerically.

```
sum( 1/n!, n=0..infinity );
```

$$e$$

```
sum( 1/(2*n+1)^2, n=0..infinity );
```

$$\frac{\pi^2}{8}$$

To compute the third infinite series numerically, we will use the **evalf** and **Sum** commands together. This format is similar to that of the numerical integration. (**Sum** is the inert command for **sum**.)

```
evalf( Sum( k/(exp(2*Pi*k) -1), k=1..infinity ));
```

$$.001877930894$$

```
evalf( 1/24 - 1/(8*Pi) );
```

$$.00187793091$$

Taylor Series

Taylor Polynomials

Recall that if a function f satisfies certain reasonable conditions, then it can be approximated by a polynomial $p_n(x)$ of degree n near a point $x = a$ defined by:

$$p_n(x) = f(a) + \frac{f'(a)}{1!}(x-a) + \frac{f''(a)}{2!}(x-a)^2 + \cdots + \frac{f^{(n)}(a)}{n!}(x-a)^n$$

The polynomial $p_n(x)$ is called the Taylor polynomial of f of degree n about $x = a$.

We can use the **add** command to write out Taylor polynomials explicitly. For example, e^{2x} has the following sixth-degree Taylor polynomial about the origin:

```
f := x -> exp(2*x):
f(0)+ eval(add(diff(f(t), t$k)*x^k/k!, k=1..6), t=0);
```

$$1 + 2x + 2x^2 + \frac{4}{3}x^3 + \frac{2}{3}x^4 + \frac{4}{15}x^5 + \frac{4}{45}x^6$$

The taylor and convert Commands

Instead of using the clumsy expression shown above, you can use the **taylor** and **convert** commands as follows to produce the Taylor polynomial of degree $n-1$ about $x = a$:

```
convert( taylor( function, x = a, n ), polynom );
```

For example,

```
convert( taylor( cos(x), x = 0, 3 ), polynom );
```

$$1 - \frac{x^2}{2}$$

```
convert( taylor( cos(x), x = 0, 8 ), polynom );
```

$$1 - \frac{x^2}{2} + \frac{x^4}{24} - \frac{x^6}{720}$$

```
convert( taylor( cos(x), x = 0.5, 7 ), polynom );
```

$$1.117295331 - .4794255386x - .4387912810(x - .5)^2$$
$$+ .07990425645(x - .5)^3 + .03656594008(x - .5)^4$$
$$- .003995212822(x - .5)^5 - .001218864669(x - .5)^6$$

The **taylor** command by itself actually gives a Taylor polynomial together with a remainder term of degree n, $O((x - a)^n)$.

```
taylor( cos(x), x = 0, 8 );
```

$$1 - \frac{x^2}{2} + \frac{x^4}{24} - \frac{x^6}{720} + O(x^8)$$

Applying **convert** with the **polynom** option to this result will drop off the remainder term and give the Taylor polynomial of degree $n-1$.

```
convert(%, polynom);
```

$$1 - \frac{x^2}{2} + \frac{x^4}{24} - \frac{x^6}{720}$$

More Examples

Compare Graphically a Function with Its Taylor Polynomials

■ **Example.** Consider the function $f(x) = e^{-0.3x} + \sin(x)$:

```
f := x -> exp(-0.3*x) + sin(x);
pict1 := plot(f(x), x=-3..3, thickness = 3) :
```

Suppose we compare f graphically with some of its Taylor polynomials, say near $x = 1$. We start with the expression of the Taylor polynomial of degree 3 (but we will not show you the picture right away).

```
g3 := convert(taylor(f(x), x=1, 4),polynom):
pict3 := plot( g3, x=-3..3 ):
```

> **Note**: We used the number 4 in the definition of the expression **g3** to give us the *third* degree polynomial.

```
with(plots):
display({pict1,pict3});      # The graph of f is thicker in the picture.
```

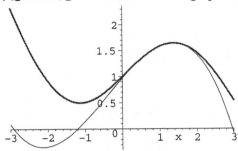

Next we increase the degree of the Taylor polynomial to 5 and then 9:

```
g5 := convert(taylor(f(x), x=1, 6),polynom):
pict5 := plot( g5, x=-3..3 ):
display({pict1,pict5});
```

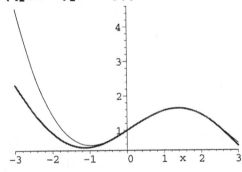

```
g9 := convert(taylor(f(x), x=1, 10),polynom):
pict9 := plot( g9, x=-3..3 ):
display({pict1,pict9});
```

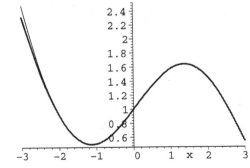

Clearly, the approximation improves as we increase the degree of the polynomial.

Interval of Convergence

■ **Example.** Consider the function $g(x) = \dfrac{1}{1+x^2}$:

```
g := x -> 1/(1+x^2):
```

The theory of infinite series tells us that the Taylor series for $g(x)$ near x = 1 only converges on a certain interval centered at 1. We would like to see this result with *Maple*. We first set up graphs of the function g and its of Taylor polynomial approximations.

```
gPict := plot(g(x), x=-2..4, y=-2..5, thickness=3):

g3 :=  convert(taylor(g(x), x=1,4),polynom):
gPict3 := plot(g3, x=-2..4, y=-2..5, color=blue):

g9 :=  convert(taylor(g(x), x=1,10),polynom):
gPict9:= plot(g9, x=-2..4, y=-2..5, color=red):

g19 :=  convert(taylor(g(x), x=1,20),polynom):
gPict19:= plot(g19, x=-2..4,y=-2..5, color=green):
```

Now we use the display command to see the graphs together.

```
with(plots):
display({gPict, gPict3, gPict9, gPict19});
```

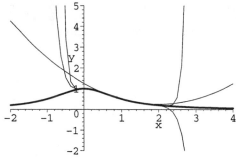

Notice that increasing the degree of the Taylor polynomial improves the approximation on a certain interval around 1. However, the polynomials turn away sharply at points outside this interval of convergence, say at $x = 2.2$. This cannot be improved no matter the degree of the Taylor polynomial (even at degree 99).

```
g99 :=  convert(taylor(g(x), x=1,100),polynom):
gPict99:= plot(g99, x=-2..4, y=-2..5, color=blue):
display({gPict, gPict99});
```

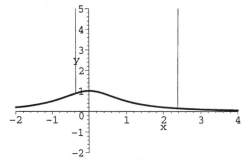

We can tell from the picture that the Taylor polynomial appears to converge to g only in the interval $-0.2 \le x \le 2.2$. One can compute theoretically that the interval of convergence is $1-\sqrt{2} \le x \le 1+\sqrt{2}$.

Troubleshooting Q & A

Question... I got some strange answers when summing a series that used names like "Psi" or "Ψ." What happened?

Answer... These are **special functions** that appear quite often in the summation of a series. Their values are well known. For example, you might see something like this:

```
sum( 1/n^2, n=1..1000 );
```

$$-\Psi(1,\ 1001) + \frac{1}{6}\pi^2$$

```
evalf(%);
```

$$1.643934568$$

Question... What is the difference in usage between the **add** and **sum** commands?

Answer... In general we like to use **add** for finite numeric sums, and use **sum** for infinite or symbol sums. (The **sum** command can also be used to compute finite numeric sums, but sometimes its approach is different from that of **add**.) *Maple* recommends the you use **add** for finite numeric sums. You can see the difference in:

```
sum( 1/n^2, n=1..1000 );
```

$$-\Psi(1,\ 1001) + \frac{1}{6}\pi^2$$

```
evalf(%);
```

$$1.643934568$$

```
add( 1/n^2, n=1..1000 );
```

(a long output due to exact representation.)

```
evalf(%);
```

$$1.643934568$$

Question... When I calculated the numerical value of an infinite series using **evalf** and **Sum**, *Maple* returned the series without evaluation. Does this mean that the series diverges?

Answer... No. This only means that *Maple* failed to give you an answer. For example, $\sum_{n=1}^{\infty} \frac{\cos(n)}{n^2}$ converges while $\sum_{n=1}^{\infty} \cos(n)$ diverges, but both will get a similar response from *Maple*, as shown below:

```
evalf( Sum(cos(n)/n^2, n=1..infinity) );
```

$$\sum_{n=1}^{\infty} \frac{\cos(n)}{n^2}$$

```
evalf( Sum(cos(n), n=1..infinity) );
```

$$\sum_{n=1}^{\infty} \cos(n)$$

Question... I found a Taylor series using the **taylor** command, but I had trouble plotting it. What should I check for?

Answer... The **taylor** command gives both a Taylor polynomial and a formal error term $O\!\left((x-a)^n\right)$. The **plot** command does not work because of this term. You have to use **convert** to drop this term before you can **plot** it. For example:

```
plot( taylor(sin(x), x=0,5), x=-3..3);
```

Plotting error, empty plot

Instead, you should use

```
plot( convert(taylor(sin(x), x=0,5), polynom),
                x=-3..3);
```

Question... When I tried plotting a function and a Taylor polynomial on the same graph, the function disappeared. What happened?

Answer... First check the plots separately. If they both work, look at the scales on the y-axis. If the scales are very different, one of the graphs may appear as a straight line over the x-axis when you put the two curves together. Look at the scale on the y-axis. The large y scale will make most interesting features of your original function disappear. Specify a range for the y-axis.

For example, the first of these is not very clear, but the second is a very nice picture:

```
plot({sin(x),
        convert(taylor(sin(x),x=0,10), polynom)},
        x= -10..10);

plot({sin(x),
        convert(taylor(sin(x),x=0,10), polynom)},
        x= -10..10, y= -2..2);
```

Question... I defined a Taylor polynomial function using the **convert** and **taylor** commands, but I have trouble evaluating the function. What should I do?

Answer... The usual way of defining a new function using **:=** does not work very well with the **convert** and **taylor** commands. For example, if you try

```
g := x -> convert(taylor(sin(x),x=0,10), polynom):
g(x);
```

$$x - \frac{1}{6}x^3 + \frac{1}{120}x^5 - \frac{1}{5040}x^7 + \frac{1}{362880}x^9$$

It looks OK, but when you ask

```
g(2);
```

Error, (in g) wrong number (or type) of parameters in function taylor

This is similar to the trouble we mentioned about the **diff** command at the end of Chapter 11. We need to use the **unapply** command in assigning the function.

```
unassign('g'):                #Clear the previous definition.
g :=
unapply(convert(taylor(sin(x),x=0,10),polynom),x);
```

$$g := x \rightarrow x - \frac{1}{6}x^3 + \frac{1}{120}x^5 - \frac{1}{5040}x^7 + \frac{1}{362880}x^9$$

```
g(x);
```

$$x - \frac{1}{6}x^3 + \frac{1}{120}x^5 - \frac{1}{5040}x^7 + \frac{1}{362880}x^9$$

```
g(2);
```

$$\frac{2578}{2835}$$

CHAPTER 14
Solving Differential Equations

Symbolic Solutions of Equations

The dsolve Command

Using the **dsolve** command, you can symbolically solve ordinary differential equations that involve $y(x)$ and x. This command is used in the form:

> **dsolve(** *the differential equation* **, y(x));**

> **Note:** The differential equation is entered using the equal sign **=** and the y must appear as **y(x)**.

Here are some simple examples.

Equation	To Solve It *in* Maple	Comment
$y' = 3x^2 y$	```dsolve(diff(y(x),x) = 3*x^2*y(x), y(x));``` $$y(x) = \_C1\, \mathbf{e}^{(x^3)}$$	**diff(y(x), x)** is used to denote y' in *Maple*.
$y\,y' = -x$	```dsolve(y(x)*diff(y(x),x) = -x, y(x));``` $$y(x) = \sqrt{-x^2 + \_C1}, \quad y(x) = -\sqrt{-x^2 + \_C1}$$	There are two solutions for this equation.
$y'' + y' - y = 0$	```dsolve(diff(y(x), x$2) + diff(y(x),x) - y(x) = 0, y(x));``` $$y(x) = \_C1\, \mathbf{e}^{\left(-\frac{1}{2}(\sqrt{5}+1)x\right)} + \_C2\, \mathbf{e}^{\left(\frac{1}{2}(\sqrt{5}-1)x\right)}$$	This is a *second*-order equation. We use **diff(y(x), x$2)** to denote y''.
$y' = -\cos(xy)$	```dsolve(diff(y(x),x) = -cos(x*y(x)), y(x));``` (no output from *Maple*)	*Maple* will not return an output if it cannot solve the differential equation.

> **Note:** $\_C1$ and $\_C2$ denote arbitrary constants in the solutions above.

Equations with Initial or Boundary Conditions

Sometimes you need to solve a differential equation subject to initial or boundary conditions. In such a case, the format of the **dsolve** command is:

> **dsolve({** *a differential equation* **,** *initial boundary condition(s)* **}, y(x));**

The table on the following page gives several examples of the use of the **dsolve** command:

Equation & Condition(s)	To Solve It in Maple
$y' = -36x$, with $y(0) = 2$	`dsolve( {diff(y(x),x) = -36*x, y(0)=2}, y(x));` $y(x) = -18x^2 + 2$
$y' = 3x^2 y$ with $y(1) = -1$	`dsolve({diff(y(x),x) = 3*x^2*y(x),y(1)=-1},y(x));` $y(x) = -\dfrac{e^{(x^3)}}{e}$

$y'' + y = 20\cos(x)$ with $y(0) = 0$, $y(\pi/2) = 1$	`dsolve( {diff(y(x),x$2) +y(x) = 20*cos(x),`
	`         y(0)=0, y(Pi/2) =1}, y(x));`
	
	$y(x) = 10\sin(x)\,x + (-5\pi+1)\sin(x)$

$y'' - 2y' + y = x$ with $y(0) = 0$, $y'(0) = 2$	`dsolve( {diff(y(x), x$2)-2*diff(y(x),x) + y(x)`
	`         = x, y(0)=0, D(y)(0)=2}, y(x));`
	
	$y(x) = 2 + x - 2e^x + 3e^x x$

> **Note:** The initial condition $y'(0) = 2$ in the last example above is entered as **D(y)(0)=2**.

Notice that in the last two examples above, two initial or boundary conditions are needed to guarantee a unique solution for a second-order differential equation.

Numerical Solutions of Equations

The numeric Option for dsolve

By adding the **numeric** option in the **dsolve** command, you can find a numerical approximation to a solution of a differential equation, subject to given initial conditions. You use it in the following format:

> **dsolve({** *differential equation, initial condition(s)***},y(x),numeric);**

For example, you can find a numerical approximation for the solution to the equation $y' = -xy$, subject to the initial condition $y(0) = 1$ with:

```
soln := dsolve( {diff(y(x),x) = -x*y(x), y(0)=1},
            y(x), numeric);
```

$soln := \mathbf{proc}(rkf45\_x) \dots \mathbf{end}$

This output looks strange. Don't worry! *Maple* reports the answer as an algorithm to approximate the solution numerically.

You can now compute, for example, the values $y(0)$, $y(0.25)$, $y(0.5)$ and $y(1)$ with:

```
[soln(0), soln(0.25), soln(0.5), soln(1)];
```

$[[x = 0,\ y(x) = 1.],\ [x = .25,\ y(x) = .969233173580462682],$
$[x = .5,\ y(x) = .882496846819249225],$
$[x = 1,\ y(x) = .606530626330481381]]$

Or you can see a graph of the solution with the **odeplot** command that is defined in the **plots** package:

```
with(plots):
```

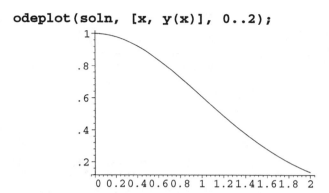

```
odeplot(soln, [x, y(x)], 0..2);
```

In general, the **odeplot** command can show the graph of the solution $y(x)$ for $a \le x \le b$ with the syntax:

```
with(plots):
odeplot( name of the result from dsolve, [x, y(x)], a..b );
```

Systems of Differential Equations

Solving a System Symbolically

If you have a system of differential equations involving functions $x(t)$ and $y(t)$, *Maple* can solve the system symbolically using the format:

```
dsolve[ { system of differential equations }, {x(t), y(t)} );
```

For example, to solve for functions $x(t)$ and $y(t)$ in the system $x'(t) = x(t) - y(t)$ and $y'(t) = y(t)$, use:

```
dsolve({diff(x(t),t) = x(t) - y(t),
            diff(y(t),t) = y(t)}, {x(t), y(t)});
```

$$\{y(t) = e^t \_C2, \quad x(t) = -e^t(-\_C1 + t\_C2)\}$$

Solving a System Numerically

The **dsolve** command can solve systems of differential equations numerically using the **numeric** option, similar to what we did earlier. For example, to solve the system $x'(t) = -2x(t)^2 - y(t)$ and $y'(t) = x(t) - y(t)$ numerically with initial conditions $x(0) = 0.2$ and $y(0) = 0.1$, use:

```
soln := dsolve({diff(x(t),t) = -2*x(t)^2 - y(t),
    diff(y(t),t) = x(t)-y(t), x(0) = 0.2, y(0) = 0.1},
    {x(t), y(t)}, numeric);
```

$$soln := \mathbf{proc}(rkf45\_x) \dots \mathbf{end}$$

Again, the output is given by an algorithm. Now, for example, you can find the values of $(x(0), y(0))$ and $(x(5), y(5))$ with:

```
[soln(0), soln(5)];
```

$$[[t = 0, x(t) = .2, y(t) = .1],$$
$$[t = 5, x(t) = -.0037309602312316638, y(t) = -.0154401644892249776]]$$

You can also see the curve $(x(t), y(t))$, called a phase diagram). (We'll draw it for for $0 \le t \le 20$ and increase the number of points to 200 to get a smoother picture.)

```
odeplot(soln, [x(t),y(t)], 0..20, numpoints = 200);
```

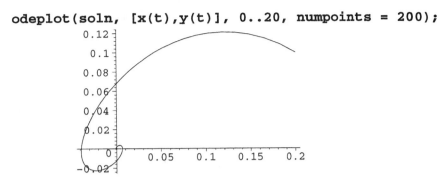

To see both graphs of $x(t)$ and $y(t)$ plotted against t, we use

```
odeplot(soln,[[t,x(t)],[t,y(t)]],0..20,numpoints=200);
```

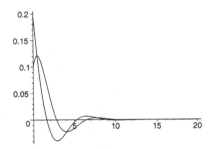

To see the solution curve in space, you can use the following (no picture shown):

```
odeplot(soln,[t,x(t),y(t)],0..20,numpoints=200);
```

More Examples

Geometry of a First-Order Differential Equation

■ **Example.** Let us find the general solution of the differential equation $y' = 3x^2 y$:

```
dsolve( diff(y(x), x) = 3*x^2*y(x), y(x));
```

$$y(x) = \_C1\ e^{(x^3)}$$

You can sketch a few particular solutions of this equation, for example, with $C1$ having values 0.5, –0.71, and 0.92. (We will not show the output here – you will see the picture later anyway):

```
curves := plot({0.5*exp(x^3), -0.71*exp(x^3),
                0.92*exp(x^3)}, x=-2..1, thickness = 3):
```

The slope field of this differential equation can be seen with the following command. We will explain this command in detail in Chapter 20 when we discuss vector fields.

```
with(plots):
pict1 := fieldplot( [1, 3*x^2*y], x=-2..2, y=-2..2 ):
```

Now you see the solutions together with the slope field. (Notice that the solution curves are tangent to the vectors that make up the slope field.)

```
display( {pict1, curves} );
```

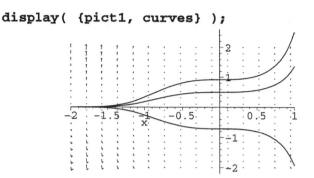

Useful Tips

💡 💡   There are library packages called **DEtools and PDETools** that include many useful commands in solving ordinary or partial differential equations.

For example, commands such as **DEplot, DEplot3d, phaseportrait, dfieldplot**, and **PDEplot** can plot the solution of a given equation (or equations), while commands such as **RiemannPsols, abelsol, bernoullisol, eulersols, matrixDE, pdsolve**, and **PdEchangecoords** may help you solve the equation.

You can load these packages using **with(DEtools);** and **with(PDEtools);** and experiment with the commands.

Troubleshooting Q & A

Question... I got an error message from **dsolve**. What should I look for?

Answer... This most likely means that you did not enter the differential equation(s) correctly. The three most common mistakes made are these:

- You did not match (*parentheses*) correctly.

- You did not follow the rules: y has to be typed as **y(x)**, y' is typed as **diff(y(x), x)**, y'' is typed as **diff(y(x), x$2)**, $y'(a) = b$ is typed as **D(y)(a) = b**, and so on.

- You assigned values earlier to either the independent variable (**x**) or the function name (**y**). Try **unassign('x', 'y');** and reexecute.

Question... When I used the **numeric** option for **dsolve,** I got an error message about "... initial conditions." What does this mean, and what should I do?

Answer... Make sure you gave the right number of initial conditions in your input. A first-order differential equation needs one initial condition, a second-order differential equation needs two initial conditions, and so on.

CHAPTER 15

Making Graphs in Space

Graphing Functions of Two Variables

The smartplot3d Command

The quickest way to sketch a surface in three dimensions (in 3-D) is to use the **smartplot3d** command.

You input an expression that gives the height of a surface above the xy-plane, in terms of the independent variables x and y. The **smartplot3d** command has the form:

> **smartplot3d(** *an expression of x and y* **);**

For example, the surface whose height is $z = 4 - x^2 - y^2$ above the xy-plane can be seen with:

```
smartplot3d( 4-x^2-y^2 );
```

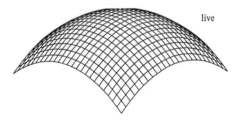

live

The plot3d Command

Although we can use **smartplot3d** to obtain a 3-D picture quickly, usually we prefer to draw pictures using the **plot3d** command. This will give us better control of the output graphics.

In the **plot3d** command, you input an expression in terms of the independent variables x and y, and specify bounds for the x- and y-variables as $x_0 \le x \le x_1$ and $y_0 \le y \le y_1$. The corresponding **plot3d** command has the form:

> **plot3d(** *an expression of x and y,* **x** = $x_0 \mathbf{..} x_1$, **Y** = $y_0 \mathbf{..} y_1$ **);**

For example, the surface whose height is $z = 4 - x^2 - y^2$ above the xy-plane, over the rectangle $-2 \le x \le 2$ and $-2 \le y \le 2$, is seen with:

```
plot3d( 4-x^2-y^2, x = -2..2, y = -2..2);
```

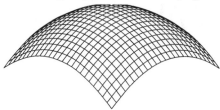

In the language of multivariable calculus, this means that **plot3d** will show you the graph of a two-variable function.

■ **Example.** The graph of the function $f(x, y) = x^2 - y^2$ looks like a saddle.

```
f := (x,y) -> x^2 - y^2 ;
plot3d( f(x,y), x = -3..3 , y = -3..3);
```

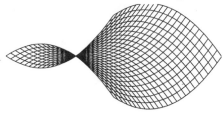

Changing the 3-D View and options from the tool bar and menu bar

Click and hold the (left) mouse button with the cursor/arrow at any point inside the 3-D picture (produced from either **smartplot3d** or **plot3d**) As you drag the mouse with the button held down, the picture rotates. This allows you to see the picture from any viewpoint.

When you click on the 3-D graphic output, the context bar changes to display a viewing angle and a series of buttons for viewing options. You can click on any of these buttons to choose the type of display you want.

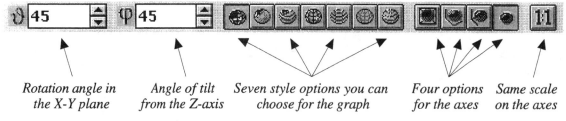

Rotation angle in the X-Y plane   *Angle of tilt from the Z-axis*   *Seven style options you can choose for the graph*   *Four options for the axes*   *Same scale on the axes*

Clicking on a 3-D graph also changes the menu bar at the top of the *Maple* window. The menus allow you to change the **Style**, **Color**, **Axes**, and **Projection** options as well as lighting.

For example, if you create a picture using **smartplot3d**, you may want to use the **Ranges** option under the **Axes** menu to adjust the intervals for the *x*, *y* and *z* axes.

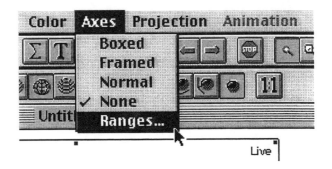

Specifying Options

The grid Option

You can use the **grid** option for **plot3d** to generate a more detailed picture. Consider:

```
plot3d( sin(3*y)+cos(5*x), x = -Pi ..Pi, y=-Pi..Pi );
```

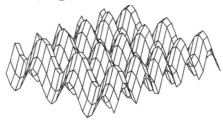

The picture does not look very good. But if we add the option **grid = [50,50]**, we'll see the graph shown with values sampled from a 50×50 grid, instead of the default 25×25 grid. This gives a smoother picture but takes longer to compute and draw.

```
plot3d( sin(3*y)+cos(5*x),
          x = -Pi ..Pi, y=-Pi..Pi, grid=[50,50] );
```

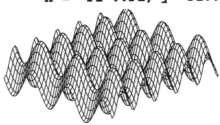

The view Option

You can use the **view** option in the **plot3d** command to control the range of the z-axis. For example, if we try the graph

```
plot3d(1/(x^2+y^2), x=-2..2, y=-2..2);
```

The result is not very informative because the values of this function close to the origin are very large and distort the scale of the z-axis. If you try,

```
plot3d(1/(x^2+y^2), x=-2..2, y=-2..2, z=-5..5);   #error
```

you will get an error message. However, you can use the **view** option as follow

```
plot3d(1/(x^2+y^2), x=-2..2, y=-2..2, view =-5..5);
```

Now you will see a nice truncated graph.

Other Useful Options

There are a number of options that you can use to add more "character" to a picture with **plot3d**. For example,

Option	What It Does
axes = normal	Shows the three axes in a normal way.
axes = boxed	Surrounds the picture with a box.
labels = ["x","y","z"]	Provides names to label each of the three axes.
scaling = constrained	Scales the graphic so that units in each direction have the same length.
color = *the color you want*	Draw the graphic with the specified color.

■ **Example.** The sombrero has equation $f(x,y) = \sin(\sqrt{x^2 + y^2})/\sqrt{x^2 + y^2}$:

```
plot3d( sin(sqrt(x^2+y^2))/sqrt(x^2+y^2), x=-7..7,
    y=-7..7,grid =[30,30], axes = boxed, color = red,
    labels=["x axis","y axis", "z axis"]);
```

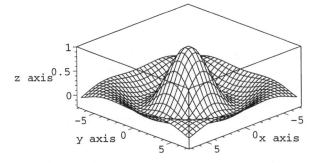

Surfaces in Cylindrical and Spherical Coordinates

Sometimes surfaces are described in terms of the cylindrical or spherical coordinate systems. *Maple* can draw such surfaces easily with the **cylinderplot** and **sphereplot** commands, respectively. These are defined in the **plots** package.

The cylinderplot Command

Points in the cylindrical coordinate system are described by quantities r, θ, and z, where

• r is the horizontal radial distance of the point from the z-axis;

• θ is the horizontal angle measured from the x-axis; and

- z is the z-coordinate in standard rectangular coordinates.

To draw the surface $r = f(\theta, z)$ for $\theta_0 \le \theta \le \theta_1$ and $z_0 \le z \le z_1$, you enter:

```
with(plots):
cylinderplot( f(θ, z), theta = θ₀..θ₁, z = z₀..z₁ );
```

> **Note:** In the **cylinderplot** command, you must enter the interval for **theta** first, then the interval for **z**. If you do not follow this order, the picture will be incorrect.

For example, to see the surface $r = z^2(\cos(3\theta))^2$, for $0 \le \theta \le 2\pi$ and $-2 \le z \le 2$:

```
with(plots):
cylinderplot( z^2*(cos(3*theta))^2, theta=0..2*Pi,
        z=-2..2);
```

Notice that this surface is very "choppy" as the horizontal angle θ varies, yet the surface is quite smooth along the vertical z-direction. This suggests that we should increase the resolution of the graphic in the θ variable (from 25 to 80) but decrease the resolution in z (from 25 to 15). We can do this with:

```
cylinderplot( z^2*(cos(3*theta))^2,
            theta=0..2*Pi, z=-2..2, grid=[80,15] );
```

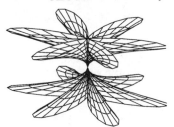

The sphereplot Command

Points in the spherical coordinate system are described by quantities ρ, θ, and ϕ, where

- ρ is the radial distance in space of the point from the origin;

- θ is the horizontal angle measured from the x-axis; and

- ϕ is the vertical angle measured from the z-axis.

To draw the surface $\rho = f(\theta, \phi)$, $\theta_0 \le \theta \le \theta_1$, and $\phi_0 \le \phi \le \phi_1$, you will enter:

```
with(plots);
sphereplot( f(θ, φ), theta = θ₀..θ₁, phi = φ₀..φ₁ );
```

> **Note:** In the **sphereplot** command, you have to enter the interval for **theta** first, then the interval for **phi**. If you do not follow this order, the picture will be incorrect.

For example, to see the surface $\rho = \sqrt{\theta}\,(3 + \cos\phi)$, $0 \le \theta \le 3\pi/2$, and $0 \le \phi \le \pi$:

```
with(plots):
sphereplot( sqrt(theta)*(3 + cos(phi)),
    theta =0..3*Pi/2, phi=0..Pi);
```

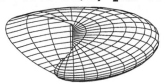

More Examples

Choosing the Right Coordinate Systems

In some cases, a surface given in rectangular coordinates will look better if you draw it using cylindrical or spherical coordinates.

■ **Example.** For example, the surface $z = \frac{x^2 - y^2}{(x^2 + y^2)^2}$ can be plotted with:

```
plot3d( (x^2-y^2)/(x^2+y^2)^2, x=-3..3, y=-3..3 );
```

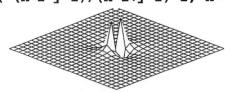

The picture is choppy especially near the origin. However, if we use cylindrical coordinates, the equation of the surface becomes

$$z = \frac{x^2 - y^2}{(x^2 + y^2)^2} = \frac{(r\cos\theta)^2 - (r\sin\theta)^2}{((r\cos\theta)^2 + (r\sin\theta)^2)^2} = \frac{r^2(\cos^2\theta - \sin^2\theta)}{r^4} = \frac{\cos 2\theta}{r^2},$$

which means $r^2 = \cos(2\theta)/z$, or equivalently $r = \sqrt{\cos(2\theta)/z}$. (We need to express the surface in the form $r = f(\theta, z)$ in order to use **cylinderplot**.)

```
top := cylinderplot( sqrt(cos(2*theta)/z),
        theta=0..2*Pi, z=0.1..2, grid =[50,15]):
bottom := cylinderplot( sqrt(cos(2*theta)/z),
        theta=0..2*Pi, z=-2..-0.1, grid =[50,15]):
display({top,bottom}, orientation=[36,77]);
```

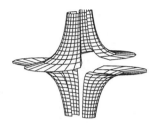

This picture is nicer!

Useful Tips

For 2-D graphs having multiple curves, *Maple* automatically uses a different color for each curve. For 3-D graphs, variation in color is used for depth. This means that if multiple surfaces are graphed together, they will have the same coloring scheme.

To designate separate colors for different surfaces in the same picture, you must plot each one separately using a different **color** option and then use the **display** command to put them together. For example:

```
pict1 := plot3d(10-x^2, x=-3..3, y=-3..3, color=red):
pict2 := plot3d(y^2, x=-3..3, y=-3..3, color=green):
with(plots):
display({pict1, pict2});
```

If you want to see the 3-D picture of an output expression quickly, you can click the output with the right mouse button (or option-key-click for Mac users). A **Context** menu pops up. You can then select **Plot** → **3D-Plot** → **x,y**. *Maple* will automatically enter the **smartplot3d** command and show you the picture.

The interactive **Plot Builder** is a useful feature with *Maple* 8 and higher. You access it in the same way you work with **smartplots**.

You click the output expression with the right mouse button (or option-key-click for Mac users). A **Context** menu pops out. You can then select **Plot** → **Plot Builder.** *Maple* will provide a series of pop up menus that walk you through the options available on the graph. When you finish, *Maple* shows the command to build the plot, then draws the graph. The plot builder can also be invoked with the interactive command from the plots package.

You can also use **plot3d** to draw parametric surface in space. We will discuss this in detail in Chapter 19.

Troubleshooting Q & A

Question... *Maple* either quit running or "crashed" when I was drawing some 3-D graphics. I also saw some messages that *Maple* "needs more memory." What should I do?

Answer... *Maple* uses a lot of memory when it creates 3-D pictures. You need a very capable system with lots of memory to run *Maple*. Most "crashes" occur because there is not enough memory on your system.

When you installed *Maple* "out of the box," your system memory parameters may not have been set high enough to sustain intensive 3-D imaging. If you can increase *Maple*'s memory partition on your system, please do so.

Question... I tried to draw a 3-D picture but got an error message telling me about a "Plotting error." What happened?

Answer... This means that *Maple* has trouble using your function you specified to generate points for the picture. Check these points:

- Did you mistype the input?
- Did you use the same variables in the function as you used in specifying the intervals?
- Is the function well-defined in the given intervals?

Another common mistake is to use the **plot** command to draw a 3-D picture; make sure you use the **plot3d** command.

Question... I drew the graph of a function using **plot3d**, but the picture looks different from the one shown in my calculus book. Why is that?

Answer... Two graphics may sometimes look different because they are drawn with different scales on the three axes. Adjust them with the **scaling** option. Also try different viewpoints.

Question... When I tried to combine various pictures from **smartplot3d** using **display**, I got an error message. Why is that?

Answer... You cannot combine the pictures from **smartplot3d** with **display**. Although the outputs from **smartplot3d** and **plot3d** appear to be similar, their structures are quite different.

The reason is that the pictures from **smartplot3d** allow several manipulations that are not offered in the standard plots. For example, you can control-drag an expression into a **smartplot3d** picture to draw that expression. As a result, we cannot combine these pictures using **display** as we can with standard plots.

Question... When I used **cylinderplot** or **sphereplot**, nothing happened. What should I do?

Answer... Did you remember to load the package **with(plots)**? Now, try the command again.

Question... The picture I got from **cylinderplot** or **sphereplot** was completely wrong. What should I check?

Answer... Check these three areas:

- Make sure you typed the input function and the intervals of the two parameters correctly.

- In **cylinderplot** you have to enter the θ-interval **theta = θ_0 .. θ_1** first, followed by the **z**-interval **z = z_0 .. z_1**. If you enter these in the wrong order, *Maple* will reverse the sense of the variables.

- In **sphereplot** you have to enter the θ-interval **theta = θ_0 .. θ_1** first, then the ϕ-interval **phi = ϕ_0 .. ϕ_1**. If you enter these in the wrong order, *Maple* will draw an incorrect picture.

Question... *Maple* either quit or "crashed" when I tried to print a worksheet that had some 3-D graphics. What should I do?

Answer... Printing 3-D graphics uses a number of software components from both your operating system and printer. The graphics format that allows live rotations on 3-D graphics conflicts with some printer drivers.

Since nice printing from *Maple* is probably not the most critical activity for you, consider a work around. Try exporting the worksheet ("**Export As**" from the file menu) as either **RTF** or as **HTML**. **RTF** stands for rich text format and is readable as a Word document. **HTML** is readable from a browser. Printing from those applications is more stable.

CHAPTER 16

Level Curves and Level Surfaces

Level Curves in the Plane

The contourplot Command

In *Maple*, the level curves (contours) of a function $f(x, y)$ are plotted with the **contourplot** command that is defined in the **plots** package. To see the level curves inside the rectangle $x_0 \le x \le x_1$, $y_0 \le y \le y_1$, you enter:

```
with(plots);
contourplot( function , x= x_0..x_1, y= y_0..y_1);
```

(This syntax looks exactly like the **plot3d** command syntax that we discussed in the previous chapter.) For example, here are some level curves of $f(x, y) = x y e^{-x^2-y^2}$:

```
contourplot(x*y*exp(-x^2-y^2), x=-2..2, y=-2..2);
```

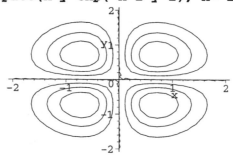

Options for ContourPlot

Some of the options we like to use with **contourplot** are the following:

Option	What It Does
contours = n	Draws n level curves
grid = [n, m]	Changes resolution of the picture
filled = true, coloring = [white, blue]	Shades the areas between level curves. Lighter blues represent lower levels, while darker blues represent higher levels.
scaling = constrained	Both the x- and y-axes are drawn with the same scale.

Here is a nicer picture than the one shown earlier:

```
contourplot(x*y*exp(-x^2-y^2), x=-2..2, y=-2..2,
    contours = 20, grid =[ 20, 20], filled=true,
    coloring=[white,blue], scaling=constrained);
```

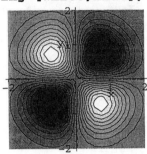

With our choice of **coloring**, lighter shades represent lower levels, while darker shades represent higher levels. We can tell from this picture that the function has its largest values near $(0.8, 0.8)$ and $(-0.8, -0.8)$.

Plotting Specific Levels

You can plot specific level curves using the **contours** option in the form **contours** = [*the levels*]. The levels must be separated by commas. For example, to see the contours at levels 0, 0.1, and 0.15, without the shading:

```
contourplot(x*y*exp(-x^2-y^2), x=-2..2, y=-2..2,
    grid = [ 30,30 ], contours = [0,0.1,0.15]);
```

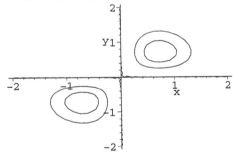

Notice that the level curve of $f(x, y) = x\,y\,e^{-x^2-y^2}$ at level 0 consists of the x- and y-axes. (*Maple* drew this level curve with squiggles near the origin.)

Level Surfaces in Space

The implicitplot3d Command

If $f(x, y, z)$ is a function of three variables defined over a rectangular region $x_0 \leq x \leq x_1$, $y_0 \leq y \leq y_1$ and $z_0 \leq z \leq z_1$, then the level surface of f at level c can be seen with the **implicitplot3d** command. It is defined in the **plots** package.

```
with(plots);
implicitplot3d( f(x, y, z)= c, x= x_0..x_1, y= y_0..y_1,
                z = z_0..z_1 );
```

You can draw more than one level surface in the same graphic, say with levels c_1 and c_2, by entering:

```
implicitplot3d({ f(x, y, z) = c₁,  f(x, y, z) = c₂ },
         x= x₀..x₁,  y= y₀..y₁,  z = z₀..z₁ );
```

Here are the level surfaces of $f(x, y, z) = x^3 - y^2 + z^2$ at the levels 1 and 10:

```
f := (x,y,z) -> x^3-y^2+z^2 ;
with(plots):
implicitplot3d({f(x,y,z)=1, f(x,y,z)=10},
      x=-2..5, y=-2..2, z=-2..3);
```

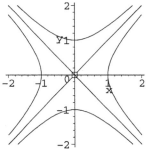

In other words, the surfaces shown above have equations $x^3 - y^2 + z^2 = 1$ and $x^3 - y^2 + z^2 = 10$.

More Examples

Comparing plot3d with contourplot

■ **Example.** Consider $f(x, y) = x^2 - y^2$. The following command will show the contours at level 0, 1, and −1.

```
contourplot(x^2- y^2, x=-2..2, y=-2..2,
      contours=[0, 1, -1], scaling=constrained) ;
```

What are these curves? Recall that a contour at level c is given by the equation $f(x, y) = c$. Thus the contour at level 0 has equation $x^2 - y^2 = 0$, the contour at level 1 has equation $x^2 - y^2 = 1$, and the contour at level −1 has equation $x^2 - y^2 = -1$. These are, respectively, the two lines $y = x$ and $y = -x$, a hyperbola that opens left/right, and a also a hyperbola that opens up/down.

We can see these contours by intersecting the graph of $f(x, y) = x^2 - y^2$ with the planes $z = 1$, $z = 0$ and $z = -1$. To make the intersection clear, we use **color** options for the individual plots, then combine them with the **display** command.

```
pict := plot3d(x^2-y^2, x=-2..2, y=-2..2,color=blue):
levela := plot3d( 1, x=-2..2, y=-2..2, color=green):
levelb := plot3d( 0, x=-2..2, y=-2..2, color=red):
levelc := plot3d( -1, x=-2..2, y=-2..2,color=yellow):

with(plots):
display({pict, levela, levelb,levelc},
            orientation =[60,70]);
```

You can also see the relationship between level curves and the graph by using the **style=patchcontour** option on the **plot3d** command. Type:

```
plot3d(x^2-y^2, x=-2..2, y=-2..2, style=patchcontour,
    contours=[0, 1, -1], axes=normal,
    orientation=[0,180] );
```

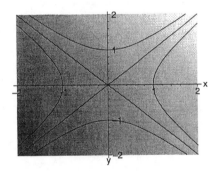

You see the level curves here. Now, click on the picture, rotate the frame box to a 3-D position, for example with an orientation of (60, 71).

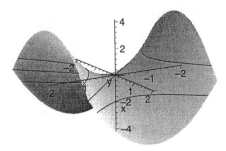

You can now see how the level curves are formed from the graph.

Curves in the Plane Defined by an Equation

■ **Example.** The equation $2x^2 - 3xy + 5y^2 - 6x + 7y = 8$ defines a rotated ellipse in the plane. We could use **implicitplot** to draw it. But it is also just the level curve of the function $f(x, y) = 2x^2 - 3xy + 5y^2 - 6x + 7y$ at level 8. We can see it with **contourplot**:

```
contourplot(2*x^2 - 3*x*y + 5*y^2 - 6*x + 7*y,
    x=-2..5, y=-3..2, contours=[8], grid=[30,30]);
```

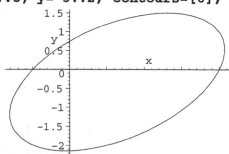

(The intervals used in the command above were arrived at after some experimentation, so that we could give you a nice picture.)

The Quadric Surfaces in Space

The Quadric Surfaces are those surfaces in space which can be given by an equation of the form:

$$Ax^2 + By^2 + Cz^2 + Dxy + Eyz + Fxz + Gx + Hy + Iz + J = 0,$$

where $A, B, C, \dots, J$ are constants. These surfaces are discussed in detail in every multivariable calculus book. With the help of the **implicitplot3d** command, we can easily see pictures of various quadric surfaces.

■ **Example.** The equation $\dfrac{x^2}{2^2} + \dfrac{y^2}{3^2} - \dfrac{z^2}{4^2} = 1$ defines a hyperboloid of one sheet:

```
implicitplot3d(x^2/2^2 + y^2/3^2 - z^2/4^2 = 1,
    x=-10..10, y=-10..10,z=-10..10 );
```

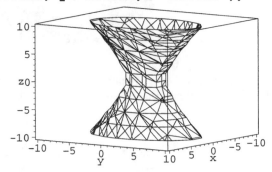

Troubleshooting Q & A

Question... When I tried **contourplot** or **implicitplot3d**, *Maple* returned the command to me unevaluated. What should I check?

Answer... Most likely you did not load the **plots** package before using the command. Type:

with(plots);

Now, try the command again.

Question... I got the error message "Plotting error, empty plot" from **contourplot**, when I specified a list of levels in the **contours** option. What happened?

Answer... There are two likely possibilities:

- The level curves you specified do not exist (e.g., the contour with level 0 is empty for the function $f(x, y) = x^2 + y^2 + 1$).

- The level curves you wanted to see do not lie inside the rectangle you gave (e.g., the contour $x^2 + y^2 = 1$ does not lie inside the rectangle $2 \le x \le 3$, $2 \le y \le 3$).

Question... I got an empty picture from **contourplot** when I specified a list of levels in the **contours** option. What happened?

Answer... Most likely the intervals you specified lie completely inside the level curve. For example, the contour $x^2 + y^2 = 9$ contains the rectangle $-2 \le x \le 2$, $-2 \le y \le 2$, so you get an empty picture when you enter:

contourplot(x^2+y^2, x=-2..2, y=-2..2,
**   contours = [9]);**

Question... I got the error message "Plotting error, empty plot" from **implicitplot3d**. What happened?

Answer... There are three likely possibilities:

- The level surface does not exist, so nothing can be shown in the output (e.g., the level surface $x^2 + y^2 + z^2 = -1$).

- The level surface you are trying to see does not lie inside the region $x_0 \le x \le x_1$, $y_0 \le y \le y_1$, and $z_0 \le z \le z_1$ you gave. (This is similar to the problem addressed in the second question above.) Recheck both the level and the region you specified.

- You did not define the three-variable function $f(x, y, z)$ correctly.

CHAPTER 17

Partial Differentiation and Multiple Integration

Partial Derivatives

The diff Command

The **diff** command we used in Chapter 11 is actually a partial differentiation operator. It differentiates an expression with respect to a specified variable, treating all other symbols as constants.

> **diff(** *the function or expression* **,** *variable* **);**

For example, if $f(x, y) = 3xy^2 - 5y \sin x$, its partial derivatives $f_x = \dfrac{\partial f}{\partial x}$ and $f_y = \dfrac{\partial f}{\partial y}$ are computed with:

```
f := (x,y) -> 3*x*y^2 - 5*y*sin(x):
diff( f(x,y), x );
```
$$3y^2 - 5y\cos(x)$$

```
diff( f(x,y), y );
```
$$6xy - 5\sin(x)$$

You find higher-order derivatives by listing the variables in the order of differentiation. For example, $f_{xxy} = \dfrac{\partial^3 f}{\partial y \partial x^2}$ represents taking the partial derivative "first by x, then by x, then by y." You compute this with:

```
diff( f(x,y), x, x, y );
```
$$5\sin(x)$$

Double and Triple Integrals

Iterated Double Integrals

The **int** command is well suited to computing iterated integrals because it treats any variable that is not named in the command as a constant. The *inner* integral in $\int_a^b \int_{g_1(x)}^{g_2(x)} f(x, y)\, dy\, dx$ is evaluated as:

```
int( f(x, y) , y = g₁(x) .. g₂(x) );
```

Then the outer integral is computed with

```
int ( % , x = a..b );
```

You can do them both at once with:

```
int(int( f(x, y), y = g₁(x)..g₂(x)), x = a..b);
```

where the inner limits are $y = g_1(x) .. g_2(x)$ and outer $x = a..b$.

Similarly, the iterated double integral $\int_c^d \int_{h_1(y)}^{h_2(y)} f(x, y)\, dx\, dy$ is computed with:

```
int(int( f(x, y), x = h₁(y)..h₂(y)), y = c..d);
```

■ **Example.** The double integral $\int_0^2 \int_0^{2x} (3x^2 + (y-2)^2)\, dy\, dx$ is

computed with:

```
int(int(3*x^2 + (y-2)^2, y=0..2*x),
x=0..2);
```

$$\frac{88}{3}$$

The integration above takes place over the region D defined by the inequalities:

$$0 \le x \le 2 \quad \text{and} \quad 0 \le y \le 2x.$$

(See the picture to the right.) D can also be described by the inequalities:

$$0 \le y \le 4 \quad \text{and} \quad y/2 \le x \le 2$$

It follows that the double integral $\int_0^2 \int_0^{2x} (3x^2 + (y-2)^2)\, dy\, dx$ has the same value

as $\int_0^4 \int_{y/2}^2 (3x^2 + (y-2)^2)\, dx\, dy$:

```
int(int(3*x^2 + (y-2)^2, x = y/2..2), y = 0..4);
```

$$\frac{88}{3}$$

Double Integrals and the VectorCalculus package The **VectorCalculus** package in *Maple* 8 and later handles multiple integrals as integration over a region rather than as an iteration of the single integral command. The example above can be integrated a number of ways, all producing the same answer we found above.

```
with(VectorCalculus):
f := 3*x^2 + (y-2)^2:

int(f, [x,y]=Region(0..2, 0..2*x));        #or
int(f, [y,x]=Region(0..4, y/2..2));        #or
int(f, [x,y]=Triangle(<0,0>,<2,0>,<2,4>));
```

Some of the other kinds of two dimensional regions recognized by the **int** command are **Circle(<*center*>,** *radius***)**, **Ellipse(***equation***)**, **Sector(Circle(<*center*>,** *radius***),** *start angle, stop angle***)**, and **Rectangle(***a..b, c..d***)**.

Iterated Triple Integrals

An iterated triple integral $\int_a^b \int_{g_1(x)}^{g_2(x)} \int_{h_1(x,\,y)}^{h_2(x,\,y)} f(x, y, z)\, dz\, dy\, dx$ is evaluated using the **int** command three times:

```
int(int(int( f(x, y, z), z = h₁(x, y)..h₂(x, y)),
         y = g₁(x)..g₂(x)), x = a..b);
```

Other variations in the order of integration can be evaluated similarly.

For example, to evaluate $\int_{-3}^3 \int_{-\sqrt{9-x^2}}^{\sqrt{9-x^2}} \int_{x+y}^{3+y} z^2\, dz\, dy\, dx$, write:

```
int(int(int(z^2, z=x+y..3+y),
       y=-sqrt(9-x^2)..sqrt(9-x^2)), x=-3..3);
```

$$\frac{567}{4}\pi$$

Or if we use the **VectorCalculus** package:

```
int( z^2, [x,y,z] = Region(
       -3..3, -sqrt(9-x^2)..sqrt(9-x^2), x+y..3+y));
```

Other three dimensional regions recognized by the **int** command of the **VectorCalculus** package include **Sphere(<center>, radius)** and **Tetrahedron(<vertex 1>, < vertex 2>, < vertex 3>, < vertex 4>)**.

Numerical Integration

If you want to find a numeric approximation for a double or triple integral, you should use **evalf** together with the **Int** commands for the outside integral and the **int** command(s) for the inside integral(s).

For example, to find $\int_{-3}^3 \int_{-\sqrt{9-x^2}}^{\sqrt{9-x^2}} \int_{x+y}^{3+y} z^2\, dz\, dy\, dx$ numerically, write:

```
evalf(Int(int(int( z^2, z = x+y..3+y),
       y=-sqrt(9-x^2)..sqrt(9-x^2)), x=-3..3));
```

445.3207587

As you may expect, numerical integration will give you an answer quickly in most cases, and it can even be used when **int** fails.

More Examples

Critical Points and the Hessian Test

■ **Example.** Suppose that $f(x, y) = x^4 - 3x^2 - 2y^3 + 3y + 0.5xy$. We can find its critical points by solving the equations $f_x = 0$ and $f_y = 0$ simultaneously:

```
f := (x,y) -> x^4 - 3*x^2 - 2*y^3 + 3*y + 0.5*x*y;
criticalpt := [solve( {diff(f(x,y),x) = 0,
                diff(f(x,y),y)=0},{x,y})];
```

$criticalpt := [\{x = -1.250162405, y = .6291421140\},$
$\{x = 1.255942703,\ y = -.7776000848\},$
$\{x = 1.191117672,\ y = .7741187286\},$
$\{x = .05935572277, y = .7105957432\},$

$$\{x = -.05877159444, \quad y = -.7036351094\},$$
$$\{x = -1.197482099, \quad y = -.6326213916\}]$$

There are six critical points. We will define the discriminant

$$D = (f_{xx})(f_{yy}) - (f_{xy})^2$$

and evaluate it at each critical point. The Hessian Test you learned in multivariable calculus says:

- If $D < 0$, the critical point is a saddle point.

- If $D > 0$, the critical point is a local maximum when f_{xx} is negative.

- If $D > 0$, the critical point is a local minimum when f_{xx} is positive.

We compute the discriminant D and f_{xx} at each of the critical points:

```
dis := diff(f(x,y),x,x)*diff(f(x,y),y,y)
                    - diff(f(x,y),x,y)^2 ;
```

$$dis := -12\,(12\,x^2 - 6)\,y - .25$$

```
eval( [dis, diff(f(x,y),x,x)], criticalpt[1] );
```

$$[-96.54552914, \quad 12.75487247]$$

Since $D < 0$, this means that the first critical point is a saddle point. We can check all the critical points at once with:

```
seq( eval( [dis, diff(f(x,y),x,x)], criticalpt[n]),
        n=1..6 );
```

$$[-96.54552914, \quad 12.75487247],$$
$$[120.3903441, \quad 12.92870488],$$
$$[-102.6671685, \quad 11.02513571],$$
$$[50.55238934, \quad -5.957722778],$$
$$[-50.56174649, \quad -5.958550796],$$
$$[84.83171040, \quad 11.20756052]$$

This result shows that the discriminant is positive at the second, fourth, and sixth of the critical points. Since f_{xx} is positive at the second and sixth, those critical points (1.256, −0.778) and (−1.197, −0.633) will be local minima for f. Also, the point (0.059, 0.711) will be a local maximum.

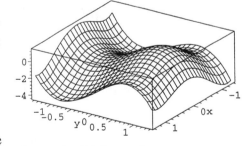

The three remaining critical points have negative discriminants, so each is a saddle point for f. You can check these points from the graph of f shown above.

Method of Lagrange Multipliers

■ **Example**. We want to find the maximum and minimum values of the function $f(x, y) = (x - 2)y + y^2$, subject to the constraint $x^2 + y^2 - 1 = 0$. The Method of Lagrange Multipliers states that we need to solve the system of equations:

$$f_x = \lambda g_x, \quad f_y = \lambda g_y \quad \text{and} \quad g = 0,$$

where g is the "constraint" function $g(x, y) = x^2 + y^2 - 1$.

```
f := (x,y) -> (x-2)*y + y^2;
g := (x,y) -> x^2 + y^2 - 1;
```

```
soln := [solve(
   { diff(f(x,y),x) = p* diff(g(x,y),x),
     diff(f(x,y),y) = p* diff(g(x,y),y), g(x,y) =0.0},
     {p, x, y} )]:
```

(Note: We use the variable **p** in this computation to stand for the multiplier λ.)

Since **soln** contains both real and complex solutions, we can use the **remove** command (as mentioned at the end of Chapter 5) to select only the real solutions.

```
realSoln := remove( has, soln, I);
```

realSoln := [{y = .7918881587, p = -.6483805940, x = -.6106661474},
{x = -.227167105, p = 2.143478901, y = -.9738557943}]]

We can evaluate the values of f at these two points with:

```
seq( eval(f(x,y), realSoln[n]), n=1..2) ;
```

$$-1.440268752, 3.117334698$$

So f has the minimum value -1.44027 at the first point $(-0.6107, 0.7919)$ and the maximum value 3.11733 at the last point $(-0.2272, -0.9739)$.

You can also see this result geometrically by drawing these two points together with the contour picture of f and the constrained circle $g = 0$:

```
pict1 := contourplot(f(x,y), x=-1.5..1.5,
              y=-1.5..1.5, contours = 20,
              filled=true, coloring=[white,blue]):

with(plots):
pict2 := implicitplot( g(x,y) = 0,
              x=-2..2, y=-2..2, thickness = 3 ):

pict3 := pointplot( [eval( [x,y], realSoln[1]),
eval([x,y], realSoln[2])],color = green, symbol =
circle, symbolsize=18):    #We will explain pointplot in Chapter 20.

display({pict1, pict2, pict3}, scaling=constrained);
```

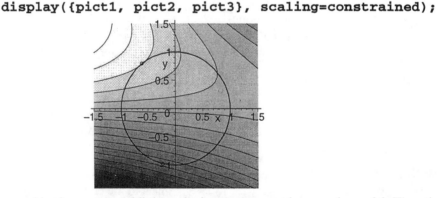

In the graphic above we used lighter shades to represent lower values of f. Thus the point $(-0.61067, 0.79189)$ on the constraint circle gives the minimum because that is where the lightest shading occurs.

Integration in Polar Coordinates

■ **Example.** Let us compute the integral $\iint_D e^{(x^2+y^2)} dA$, where D is the circular sector given in polar coordinates as $0 \le r \le 1$, $\pi/4 \le \theta \le \pi/2$. ($D$ is sketched to the right.)

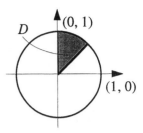

We learned from calculus that we must first write the integral as the iterated integral $\int_{\pi/4}^{\pi/2} \int_0^1 e^{r^2} r \, dr \, d\theta$. Then we can compute its value:

```
int(int(exp(r^2)*r, r=0..1), t=Pi/4..Pi/2);
```

$$\frac{1}{8} \mathbf{e} \pi - \frac{1}{8} \pi$$

Outsmarting Maple

■ **Example.** You can outsmart *Maple* easily. Let D be the region inside the circle $x^2 + y^2 = 1$. Find the double integral $\iint_D \sqrt[3]{x^2 + y^2} \, dA$ both by hand and using the computer.

- For the computer: Evaluate the double integral using rectangular coordinates. The region D is described by the inequalities $-1 \le x \le 1$ and $-\sqrt{1-x^2} \le y \le \sqrt{1-x^2}$, so you enter:

```
int(int((x^2+y^2)^(1/3),
        y = -sqrt(1-x^2)..sqrt(1-x^2)), x=-1..1);
```

After a while, *Maple* will give you a complicated answer involving hypergeometric functions and the gamma function Γ.

- However, you can evaluate the integral directly and easily using polar coordinates, because the region D is described by the inequalities $0 \le r \le 1$ and $0 \le \theta \le 2\pi$:

$$\iint_D \sqrt[3]{x^2 + y^2} \, dA = \int_0^{2\pi} \int_0^1 (r^2)^{1/3} r \, dr \, d\theta = \int_0^{2\pi} \int_0^1 (r^{5/3}) \, dr \, d\theta$$

$$= 2\pi \left(\tfrac{3}{8} r^{8/3} \Big|_0^1 \right) = \tfrac{3\pi}{4}.$$

You Win!!!

Is Fubini's Theorem wrong?

■ **Example.** You may recall that for an iterated integral, the order of integration should not affect the value of the double integral. This is called Fubini's Theorem, which basically says that $\int_a^b \int_c^d f(x,y) \, dy \, dx = \int_c^d \int_a^b f(x,y) \, dx \, dy$.

However, if you try:

```
int(int( (x+y)/(x-y)^3, y=-1..0 ), x=0..1);
```

$$\frac{1}{2}$$

```
int(int( (x+y)/(x-y)^3, x=0..1 ), y=-1..0);
```

$$-\frac{1}{2}$$

This is because the function $\dfrac{x+y}{(x-y)^3}$ is not continuous along the line $y=x$, we cannot apply Fubini's Theorem and changing the order of integration may affect the result of the integral.

Useful Tips

👁 👁   When double or triple integrals are complicated, it is easy to mistype the integrand or forget to match parentheses correctly.

We recommend that you first make a copy of the integral and use the inert **Int** command so that you can see the integral you are trying to solve. Then, once you have the right syntax, replace the **Int** commands by **int**.

For example:

```
Int(Int(exp(r^2)*r, r=0..1), t=Pi/4..Pi/2);
```

$$\int_{\frac{1}{4}\pi}^{\frac{1}{2}\pi} \int_0^1 e^{r^2} \, r \, dr \, d\theta$$

```
int(int(exp(r^2)*r, r=0..1), t=Pi/4..Pi/2);
```

$$\frac{1}{8} e \pi - \frac{1}{8} \pi$$

👁 👁   When you use the **solve** command in finding the critical points or the Lagrange multiplier method, *Maple* may give you the answers in the form "RootOf". You can avoid this by adding a decimal place to any one of the integer coefficients of the equations. (In our method of Lagrange Multiplier example earlier, we replace "0" with "0.0" in the equation.) *Maple* will then give you the answers in numerical form.

Troubleshooting Q & A

Question... I entered a double (or triple) integral expression and have been waiting for a response for two or three minutes. *Maple* has not given me an answer yet. Is something wrong?

Answer... Integration is a difficult mathematical problem. Unfortunately, *Maple* cannot solve every integration problem you encounter.

You may have to abort your calculation and look for ways to simplify the integrand (e.g., through use of cylindrical or spherical coordinates in three variables) in order to move toward getting a result.

A slow computer often takes several minutes to evaluate a complicated double or triple integral. You may simply need either a faster computer or more patience.

However, if a numerical answer is acceptable, you can replace the outer integral with **evalf** and **Int** and *Maple* will probably give you an answer very quickly.

Question... *Maple* did not give me a number when I evaluated a double or triple integral, but I know the answer should be numerical. What should I look for?

Answer... Check for these problem areas:

- It could be a mathematical error. Did you set up the integral correctly? Recheck your limits of integration, too.

- It could be a *Maple* input error. Did you type the integrand correctly? Did you input the order of integration correctly?

- The function may have variables that are not involved in the integration and that do not evaluate as numbers. For example, your input function might be written in terms of x and y, but you are doing the integration with respect to polar coordinates **theta** and **r**.

CHAPTER 18

Matrices and Vectors

Vectors

Column and Row Vectors

In *Maple* we define column vectors with the syntax < *element1* , *element2* , … >. Row vectors are defined by separating the entries with the vertical bar symbol "|", and appear in the form < *element1* | *element2* | … >.

For example, the column vector $\begin{pmatrix} 1 \\ 2 \end{pmatrix}$ and the row vector (3 4 5) are entered as

    <1, 2>;   #column vector

$$\begin{bmatrix} 1 \\ 2 \end{bmatrix}$$

    <3 | 4| 5>;   #row vector

$$[3, 4, 5]$$

You do addition, subtraction, and scalar multiplication of vectors directly using the operations **+**, **–** and *****.

    <1, 1, 2> + <2, -1, 3>;

$$\begin{bmatrix} 3 \\ 0 \\ 5 \end{bmatrix}$$

    <1| 2| 3> - 3*<2| 1| 1>;

$$[5, -1, 0]$$

Dot Product and Cross Product

You compute the dot product of two vectors by typing a period "**.**" (literally, a "dot"!):

    <1, 2, 3> . <2, 1, 1>;

$$7$$

You can also compute the dot product of two vectors with the **DotProduct** command, which is defined in the **LinearAlgebra** package.

```
with(LinearAlgebra):
DotProduct(<a1, a2, a3>, <b1, b2, b2>);
```

$$\overline{a1}b1 + \overline{a2}b2 + \overline{a3}b3$$

(Note that *Maple* uses the complex inner product by default.)

To compute a cross product of two 3-D vectors (also known as the vector product), you use the **CrossProduct** command or use the **&x** operator. Both are defined in the **LinearAlgebra** package:

```
with(LinearAlgebra):
CrossProduct(<a1| a2| a3>, <b1| b2| b3>);   #or
<a1| a2| a3> &x <b1| b2| b3> ;          #for Maple 8
```

$$[a2b3 - a3b2, a3b1 - a1\,b3, a1\,b2 - a2b1]$$

Matrices

Defining Matrices

You define a matrix in *Maple* by using the **Matrix** command. For example, to enter the matrix $\begin{pmatrix} 3 & -4 & 7 \\ -1 & 0 & 5 \end{pmatrix}$, you type:

```
Matrix([ [3,-4,7], [-1,0,5] ]);
```

$$\begin{bmatrix} 3 & -4 & 7 \\ -1 & 0 & 5 \end{bmatrix}$$

In general, a matrix is entered in the form:

Matrix([*list of row1 entries* , *list of row2 entries* , *etc.* **]);**

Notice that the above matrix can also be thought of as a column of two row vectors or as a row of three column vectors. We can also define the matrix in the following short-cut using column vectors and row vectors.

```
< <3| -4| 7>,   <-1| 0| 5> >;      #column of 2 row vectors

< <3, -1> | <-4, 0> | <7, 5> >; # row of 3 column vectors
```

$$\begin{bmatrix} 3 & -4 & 7 \\ -1 & 0 & 5 \end{bmatrix}$$

> **Note:** The **Matrix** command needs both the **[** *square brackets* **]** around the lists and **(** *parentheses* **)** at the outer level of the command. The shortcut construction uses **<** *angle brackets* **>**.

Basic Operations Involving Matrices

You can do matrix addition and subtraction, scalar multiplication and scalar addition (adding a scalar to the main diagonal), multiplication of a matrix by a matrix or vector, and power of a square matrix by using the standard operators **+**, **−**, *****, **.** (dot), and **^**. Here are a few examples.

Operation	Example	Maple *Command*

+
 Addition
	$\begin{pmatrix} 3 & -4 & 7 \\ -1 & 0 & 5 \end{pmatrix} + \begin{pmatrix} 1 & 2 & 3 \\ 4 & 5 & 6 \end{pmatrix}$	`<<3|-4|7>,<-1|0|5> +`
		 `  <<1|2|3>,<4|5|6>;`
		 $\begin{bmatrix} 4 & -2 & 10 \\ 3 & 5 & 11 \end{bmatrix}$

Operation		Maple Command and Example
– Subtraction	$\begin{pmatrix} 3 & -4 \\ -1 & 0 \\ 7 & 5 \end{pmatrix} - \begin{pmatrix} 1 & 2 \\ 3 & 4 \\ 5 & 6 \end{pmatrix}$	`<<3,-1,7>	<-4,0,5>> -` ` <<1,3,5>	<2,4,6>>;` $\begin{bmatrix} 2 & -6 \\ -4 & -4 \\ 2 & -1 \end{bmatrix}$
* Scalar multiplication	$5\begin{pmatrix} 3 & -4 & 7 \\ -1 & 0 & 5 \end{pmatrix}$	`5*<<3|-4|7>,<-1|0|5>>;` $\begin{bmatrix} 15 & -20 & 35 \\ -5 & 0 & 25 \end{bmatrix}$
+ Scalar Addition	$4 + \begin{pmatrix} 3 & -4 & 7 \\ -1 & 0 & 5 \end{pmatrix}$	`4+<<3|-4|7>,<-1|0|5>>;` $\begin{bmatrix} 7 & -4 & 7 \\ -1 & 4 & 5 \end{bmatrix}$

.
Matrix
multiplication
	$\begin{pmatrix} 3 & -4 & 7 \\ -1 & 0 & 5 \end{pmatrix} \cdot \begin{pmatrix} 2 & 0 & 1 \\ -1 & 1 & 2 \\ 3 & 5 & -2 \end{pmatrix}$	`a := <<3|-4|7>, <-1|0|5>>:`
		`b:=<<2,-1,3>|<0,1,5>|`
		`      <1,2,-2>>:`
		`a.b;`
		$\begin{bmatrix} 31 & 31 & -19 \\ 13 & 25 & -11 \end{bmatrix}$

^ Matrix Power	$\begin{pmatrix} 1 & 2 \\ 3 & 4 \end{pmatrix}^5$	`<<1,3>|<2,4>>^5;` $\begin{bmatrix} 1069 & 1558 \\ 2337 & 3406 \end{bmatrix}$

Some Useful Matrix Commands

Maple has several commands that let you easily find the inverse, the determinant, the eigenvalues, and the eigenvectors of a given square matrix. All of these commands are defined in the **LinearAlgebra** package.

Here is a quick summary:

with(LinearAlgebra):   # Use if you have not loaded the package yet.

Operation	Maple *Command and Example*

MatrixInverse
– Find the inverse of a square matrix.
	`A := Matrix([[2,5,1],[3,1,2],`
	`            [-2,1,0]]):`
	`MatrixInverse(A);`
	$\begin{bmatrix} \frac{2}{19} & \frac{-1}{19} & \frac{-9}{19} \\ \frac{4}{19} & \frac{-2}{19} & \frac{1}{19} \\ \frac{-5}{19} & \frac{12}{19} & \frac{13}{19} \end{bmatrix}$

Determinant
– Find the determinant of a square matrix.
	`B := Matrix([[2,5,1], [3,1,2], [-2,1,0]]):`
	`Determinant(B);`
	-19

Eigenvalues
– Find the eigenvalues of a square matrix.
	```C := Matrix([[1,2,-1], [2,3,1], [1,0,2]]):```
	```Eigenvalues(C);```
	
	$$\begin{bmatrix} 1 \\ \dfrac{5}{2} + \dfrac{1}{2}\sqrt{13} \\ \dfrac{5}{2} - \dfrac{1}{2}\sqrt{13} \end{bmatrix}$$

Eigenvectors
– Find the eigenvectors of a square matrix.
	```C := Matrix([[2,1,0], [-1,0,1], [1,3,1]]):```
	```Eigenvectors(C, output = list);```
	
	$$\left[\left[-1,\ 1,\ \left\{\begin{bmatrix}1\\-3\\4\end{bmatrix}\right\}\right],\ \left[2,\ 2,\ \left\{\begin{bmatrix}1\\0\\1\end{bmatrix}\right\}\right]\right]$$
	
	This means that the eigenvalue –1 has multiplicity 1 with eigenvector (1, –3, 4); eigenvalue 2 has multiplicity 2 with eigenvector (1, 0, 1);

Elementary Row Transformations

Converting Equations to Matrices

A major use of linear algebra is to solve systems of linear equations. To do that you convert the system of equations into an augmented matrix, then row reduce the matrix, then convert back to a system of equations. The conversion to a matrix is done with the **GenerateMatrix** command in the **LinearAlgebra** package.

```
GenerateMatrix([ list of equations,  list of variables,
                                        augmented=true]);
```

```
eqns := [2*x + y - z = 1, x + 3*z = 4,
            -5*x -3*y + z = 2]:
vars := [x, y, z]:
with(LinearAlgebra):
A1 := GenerateMatrix(eqns, vars, augmented=true);
```

$$A1 := \begin{bmatrix} 2 & 1 & -1 & 1 \\ 1 & 0 & 3 & 4 \\ -5 & -3 & 1 & 2 \end{bmatrix}$$

Row Operations

Maple can perform row transformations using the **RowOperation** command in the **LinearAlgebra** package, as shown below:

Row Transformation	Maple *Command*
Interchange row i with row j in matrix A.	```with(LinearAlgebra):``` ```RowOperation(A, [i, j]);```
Multiply row i by m in matrix A.	```RowOperation(A, i, m);```
In matrix A, replace row "i" by "row $i + m$ *row j."	```RowOperation(A, [i, j], m);```

■ **Example.** Consider the augmented matrix $\begin{pmatrix} 2 & 1 & -1 & 1 \\ 1 & 0 & 3 & 4 \\ -5 & -3 & 1 & 2 \end{pmatrix}$ that we obtained above.

```
with(LinearAlgebra):
A1 := Matrix([[2,1,-1,1], [1,0,3,4], [-5,-3,1,2]]) ;
```

$$A1 := \begin{bmatrix} 2 & 1 & -1 & 1 \\ 1 & 0 & 3 & 4 \\ -5 & -3 & 1 & 2 \end{bmatrix}$$

Step 1: We would like to interchange Row 1 and Row 2.

```
A2 := RowOperation(A1, [1,2]);
```

$$A2 := \begin{bmatrix} 1 & 0 & 3 & 4 \\ 2 & 1 & -1 & 1 \\ -5 & -3 & 1 & 2 \end{bmatrix}$$

Step 2: We would like to replace Row 2 by "Row 2 – 2*Row 1."

```
A3 := RowOperation(A2, [2, 1], -2);
```

$$A3 := \begin{bmatrix} 1 & 0 & 3 & 4 \\ 0 & 1 & -7 & -7 \\ -5 & -3 & 1 & 2 \end{bmatrix}$$

Step 3: We would like to replace Row 3 by "Row 3 + 5*Row 1."

```
A4 := RowOperation(A3, [3, 1], 5):
```

Step 4: Replace Row 3 by "Row 3+3*Row 2."

```
A5 := RowOperation(A4, [3,2], 3):
```

Step 5: Multiply Row 3 by –1/5.

```
A6 := RowOperation(A5, 3, -1/5);
```

$$A6 := \begin{bmatrix} 1 & 0 & 3 & 4 \\ 0 & 1 & -7 & -7 \\ 0 & 0 & 1 & \frac{-1}{5} \end{bmatrix}$$

We have transformed the original matrix into a simpler form corresponding to an equivalent linear system. This is called the **row echelon form** of a matrix in linear algebra.

To complete the Gauss-Jordan elimination, we need to:

Steps 6 and 7: Replace Row 1 by "Row 1 – 3*Row 3" and replace Row 2 by "Row 2 + 7*Row 3."

```
A7 := RowOperation(A6, [1,3], -3):
A8 := RowOperation(A7, [2, 3], 7);
```

$$A8 := \begin{bmatrix} 1 & 0 & 0 & \frac{23}{5} \\ 0 & 1 & 0 & \frac{-42}{5} \\ 0 & 0 & 1 & \frac{-1}{5} \end{bmatrix}$$

Bingo! This is called the **reduced row echelon form** of a matrix in linear algebra.

Converting Matrices to Equations

Once we obtain the reduced row echelon form, we can use the **GenerateEquations** command (inside the **LinearAlgebra** package) to convert the matrix back to the equations and see the answer.

```
with(LinearAlgebra):
GenerateEquations(A8, [x,y,z]);
```

$$\left[x = \frac{23}{5}, y = \frac{-42}{5}, z = \frac{-1}{5} \right]$$

Useful Tips

💡💡💡💡 *Maple* has two packages to work with linear algebra, the **LinearAlgebra** package (introduced in *Maple* 6) we discussed in this chapter and the **linalg** package (used by earlier versions but still active). The two packages use different data types for vectors and matrices. Also, the older package tends to abbreviate names and use lower case while the upper package uses full words and upper case first letters.

For example, corresponding to the **Linear Algebra**'s **DotProduct** and **CrossProduct** commands are **linalg**'s **dotprod** and **crossprod** commands. It is probably easiest to only use one package if possible to avoid confusion. For more information on the two packages, see the overview pages in the help material.

```
?LinearAlgebra;
?linalg;
```

💡💡💡 The **LinearAlgebra** package contains commands for construction of a number of special matrices and vectors. The commands **BandMatrix**, **ConstantMatrix**, **DiagonalMatrix**, **IdentityMatrix**, **RandomMatrix**, **RandomVector**, **ScalarMatrix**, **ScalarVector**, **UnitVector**, **ZeroMatrix**, and **ZeroVector** create the kinds of matrices and vectors suggested by the command names.

💡💡 The **LinearAlgebra** package defines many other useful commands. Some of the useful commands are **RowSpace**, **ColumnSpace**, **NullSpace**, **JordanForm**, **RowReducedEchelonForm**, **Adjoint**, **SmithForm**, **LUDecomposition**, **QRDecomposition**, **Minor**, **MinimumPolynomial**, and **SubMatrix**.

Troubleshooting Q & A

Question... When I tried to create a matrix with the angle bracket shortcut I got an error message "`... is too tall or too short`" or "`... this entry is too wide or too narrow`" or "`...parameter is non-conformant`". What should I check for?

Answer... Check that the rows or columns all have the same number of entries.

Question... When I tried to create a matrix with the **Matrix** command, the result had extra zeroes that I did not enter. How did this happen?

Answer... Check that the rows or columns all have the same number of entries. The **Matrix** command pads short rows out with zeroes to give a rectangular array.

Question... When I used the **MatrixInverse** or **RowOperation** command, *Maple* returned the command unevaluated. What should I check for?

Answer... Check that you load the package

```
with(LinearAlgebra):
```

Question... When I tried to add, subtract, or multiply two matrices, I got an error message. What should I check for?

Answer... Here are some suggestions:

- First check that the dimensions on the matrices are compatible with the indicated operations.

- Second check that the input matrices are of the correct data type. Are they generated by the **Matrix** command for the **LinearAlgebra** package or by the **matrix** command for the **linalg** package? The command

```
type(A, Matrix);
```

can check if the matrix *A* is of the data type for **LinearAlgebra** package. If the data type is incorrect, it can be fixed with

```
convert(A, Matrix);
```

Question... When I used the **MatrixInverse** or **RowOperation** command, I got an error message. What should I check for?

Answer... Same answer as above.

Question... When doing a row reduction, the example in this chapter named each of the intermediate results. Wouldn't it be easier to use **%** at each step to refer to the previous output?

Answer... Yes, if you have a lot of confidence that you will never make a mistake when you type. As noted in Chapter 2 when we first introduced **%**, this symbol should be used sparingly in long computations. If you make a mistake in the 8th step of the row reduction, you only need to retype the last statement using the result of the 7th step, while using the **%** symbol may mean that you have to go back to execute all the previous row transformations from the beginning!!

CHAPTER 19

Parametric Curves and Surfaces in Space

Parametric Curves in Space

The spacecurve Command

You use the **spacecurve** command, defined in the **plots** package, to draw a parametric curve in space. To see the curve given as $(x(t), y(t), z(t))$, for $a \leq t \leq b$, type:

```
with(plots):
spacecurve( [ x(t) , y(t) , z(t) , t = a..b ] );
```

This format is similar to the syntax of the **plot** command used for plane parametric curves. Here, however, the curve is defined with *three* parametric functions rather than *two*.

■ **Example.** The helix given parametrically by $(t, 3\cos(t), 3\sin(t))$, for $0 \leq t \leq 8\pi$, is drawn with:

```
with(plots):
spacecurve( [t, 3*cos(t), 3*sin(t), t = 0..8*Pi] );
```

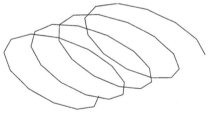

Options for spacecurve

Most options you can use with **plot3d** such as **axes**, **labels**, **color**, and **scaling** also work with **spacecurve**. By specifying a value for the **numpoints** option, you can control the resolution of the picture you see. (The default value is 50.) Here is a nicer picture of the same helix.

```
spacecurve( [t, 3*cos(t), 3*sin(t), t = 0..8*Pi],
        axes = normal,
        scaling = constrained,
        color = black,  thickness= 3,
        numpoints = 100);
```

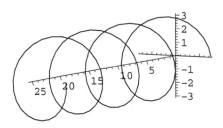

Parametric Surfaces in Space

**The plot3d
Command**

The **plot3d** command that we used to draw the graph of a two-variable function in
Chapter 15 can also be used to draw a parametric surface in space. If a surface is
defined parametrically by $(x(u,v), y(u,v), z(u,v))$ for $u_0 \leq u \leq u_1$ and $v_0 \leq v \leq v_1$,
you enter:

```
plot3d([ x(u, v) , y(u, v) , z(u, v) ],
                 u = u0..u1, v = v0..v1);
```

■ **Example.** To see a portion of the one-sheeted hyperboloid given parametrically
by $(\cos(u)\cosh(v), \sin(u)\cosh(v), \sinh(v))$, for $0 \leq u \leq 2\pi$ and $-2 \leq v \leq 2$, you
write:

```
plot3d( [cos(u)*cosh(v),sin(u)*cosh(v),sinh(v)],
           u = 0..2*Pi, v = -2..2);
```

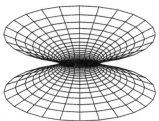

The surface $(v(2-\cos(4u))\cos(u), v(2-\cos(4u))\sin(u), v^2)$, for $0 \leq u \leq 2\pi$ and
$0 \leq v \leq 2$, gives a very nice picture of a vase:

```
plot3d(
  [v*(2-cos(4*u))*cos(u),v*(2-cos(4*u))*sin(u),v^2],
       u = 0..2*Pi, v = 0..2,
       grid = [60, 30]);
```

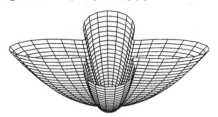

Plotting Multiple Curves and Surfaces

The **spacecurve** (or **plot3d**) command can sketch several curves (or surfaces) with a single command, just as you have already seen in the **plot**, **polarplot**, and **implicitplot** commands. Use one of these formats:

```
spacecurve({ [curve1, t=a₁..b₁], [curve2, t=a₂..b₂]});
```

or

```
plot3d({ [surface1 ], [surface2 ]}, u = u₀..u₁, v = v₀..v₁);
```

> **Note:** In the **spacecurve** and **plot3d** commands, multiple curves or surfaces can only be grouped together using **{** *curly braces* **}**, and not **[** *square brackets* **]**. This is totally different from the format of the **plot** command.

Multiple Curves

■ **Example.** Consider the helixes $(t, -3\sin(t), 3\cos(t))$ with $\pi \le t \le 6\pi$ and $(t, 3\cos(t), 3\sin(t))$ with $0 \le t \le 8\pi$. They can be drawn together using:

```
spacecurve( { [t,-3*sin(t),3*cos(t), t=Pi..6*Pi],
              [t,3*cos(t),3*sin(t), t=0..8*Pi] },
            numpoints = 100, color = black);
```

Multiple Surfaces

■ **Example.** The paraboloid $(r\cos t, r\sin t, 2 - r^2)$ opens down, while the paraboloid $(r\cos t, r\sin t, r^2)$ opens up. We can combine them to form a nice "beehive."

```
plot3d( { [r*cos(t),r*sin(t),r^2],
          [r*cos(t),r*sin(t), 2-r^2] },
        r = 0..1, t = 0..2*Pi);
```

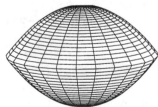

Shading and Coloring

Shading for Surfaces

The **plot3d** command illuminates a surface according to a default lighting scheme. This causes different portions of the surface to be colored or shaded differently.

You can turn off *Maple*'s default shading by setting **shading = none**. This will give an "opaque, plain" image. (Most 3-D pictures in this book are shown this way to present a clearer figure.) For example:

```
plot3d( [2*cos(t),2*sin(t),z], t = 0..2*Pi, z=-4..4,
        scaling = constrained, shading=none);
```

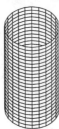

You can also control the shading more directly by specifying, for example, **shading = zgreyscale**. This will shade the cylinder above according to its height, with darker shading at the bottom:

```
plot3d( [2*cos(t),2*sin(t),z], t = 0..2*Pi, z=-4..4,
    scaling = constrained, shading=zgreyscale,
    style = patch);
```

Coloring Function

You can explicitly set the coloring of points on a surface by using the **color** option and the HUE coloring scheme. For example,

```
s := (u,v) -> [u^2, v, v^3];
plot3d( s(u,v), u=-2..2, v=-2..2, style = patch,
             color = [(u+v)/4, u^2/4, sin(v)^2]);
```

This means that at each point $s(u,v)$, it is painted with the color defined by $\text{HUE}\left[\frac{u+v}{4}, \frac{u^2}{4}, \sin^2(v)\right]$. For example, at $s(0,1)$ the color is $\text{HUE}[1/4, 0, \sin^2 1]$.

More Examples

Combining Graphics with the display Command

You can combine curves, surfaces, and other three-dimensional images into a single graphic using the **display** command (just as we saw in Chapter 15).

■ **Example.** The upper hemisphere of the unit sphere $x^2 + y^2 + z^2 = 1$ is given by ($r\cos(t)$, $r\sin(t)$, $\sqrt{1-r^2}$), for $0 \le r \le 1$ and $0 \le t \le 2\pi$.

```
g1 := plot3d( [r*cos(t), r*sin(t), sqrt(1-r^2)],
              r = 0..1, t = 0..2*Pi):
```

The point $P = (\frac{1}{2}, \frac{1}{2}, \frac{1}{\sqrt{2}})$ lies on this hemisphere. A normal (perpendicular) line to the hemisphere at P is given by $\bar{r}(t) = (\frac{1}{2}+t, \frac{1}{2}+t, \frac{1}{\sqrt{2}}+\sqrt{2}\,t)$. This command shows just a portion of the normal line:

```
with(plots):
g2 := spacecurve( [1/2+t,1/2+t, 1/sqrt(2)+sqrt(2)*t],
              thickness = 3, t=0..0.15):
```

Finally, the plane tangent to the hemisphere at P has equation $z = (2 - x - y)/\sqrt{2}$. You can sketch a portion of it near the point P with:

```
g3 := plot3d( (2-x-y)/sqrt(2),
              x = 0.2..0.8, y = 0.2..0.8, grid = [2,2] ):
```

You can now see one of the nicest features of *Maple*, the ability to combine these graphics, despite the fact that each was drawn using a different type of command.

```
display({g1,g2,g3}, orientation = [97,78],
              scaling = constrained);
```

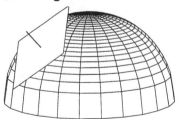

If you rotate the 3-D graph, you will see that we have described both the tangent plane and the normal line correctly.

Tangent-Normal-Binormal Frame

■ **Example.** Consider the curve $(\sin t, \ t/4, \ \cos t)$.

```
curve1:=spacecurve([sin(t), t/4, cos(t)], t=0..3*Pi,
              color = black, thickness = 2, axes=boxed):
```

We can find its tangent vector, normal vector, and binormal vector using the **TangentVector**, **PrincipalNormal** and **Binormal** commands inside the **VectorCalculus** package. (The **VectorCalculus** package is not available for versions earlier than *Maple* 8.) Then we can use the **arrow** command (from the **plots** package) to draw these vectors at various points of the curve. One can then see how the positions of these vectors are related to the curve.

```
with(VectorCalculus):
c1 := <sin(t), t/4, cos(t)>:     #you need to describe the
                                  #curve as a parameterized vector
tvplot := arrow({seq(
    eval([c1,TangentVector(c1,t)],t=i*Pi/2), i=1..5)},
    color = blue):
nvplot := arrow({seq(
    eval([c1,PrincipalNormal(c1,t)],t=i*Pi/2,
        i=1..5)}, color = red):
bvplot := arrow({seq(
    eval([c1,Binormal(c1,t)], t=i*Pi/2), i=1..5)},
    color = brown):

display({ curve1, tvplot, nvplot, bvplot},
        axes = boxed, scaling = constrained);
```

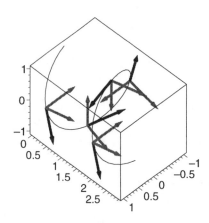

Arc Length, Line Integrals, and Surface Integrals

■ **Example.** i) Consider the curve $(t, 3\cos t, 3\sin t)$, $0 \le t \le 8\pi$. We can find the length of the curve in space with the **ArcLength** command.

```
with(VectorCalculus):
C := <t,3*cos(t),3*sin(t)>:
ArcLength(C, t=0..8*Pi);
```

$$8\sqrt{10}\pi$$

The arc length of a curve can also be obtained by integrating the value 1 along the path, and this is easily done with the **PathInt** command.

```
PathInt(1, [x,y,z]=Path(C, t=0..8*Pi));
```

$$8\sqrt{10}\pi$$

The **PathInt** command can be used to integrate other functions over a path.

```
PathInt(x^2+y, [x,y,z]=Path(C, t=0..8*Pi));
```

$$\frac{512\sqrt{10}\pi^3}{3}$$

ii) The **SurfaceInt** command in the **VectorCalculus** package is the analog of the **PathInt** command for integrating an expression over a surface.

To find the surface area of a sphere of radius r and center at (a, b, c), compute the surface integral of 1:

`SurfaceInt(1,[x,y,z]=Sphere(<a,b,c>, r));`

$$4r^2\pi$$

To integrate $x + y^2 + z^3$ over the same sphere, change the 1st argument of the command above.

`SurfaceInt(x+y^2+z^3,[x,y,z]=Sphere(<a,b,c>, r));`

$$4r^2 a\pi + 4r^2 b^2\pi + \frac{4r^4\pi}{3} + 4r^2 c^3\pi + 4r^4 c\pi$$

You can find more discussion of line and surface integrals in Chapter 20.

Troubleshooting Q & A

Question... When I tried to use the **spacecurve** command, *Maple* returned the input unevaluated. What went wrong?

Answer... You forgot to load the **plots** package first. Type:

`with(plots);`

Question... When I tried to draw a parametric curve or surface in space, I got an error message. What should I check?

Answer... There are many possibilities, but our best suggestions are:

- Did you use the **spacecurve** or **plot3d** command? A common mistake is trying to draw a 3-D picture with the **plot** command.

- Check that you have followed the correct syntax of entering the commands. The formats for **spacecurve** and **plot3d** are different and can cause confusion.

- Did you enter the function(s) correctly without a typing mistake? Is each of the functions defined everywhere in the interval(s) you specified?

- Did you use the same literal parameter in both your function and interval? (For example, check that you didn't write **f(x,y)** when the parameters were **u** and **v**.)

- Did you use two parameters for a surface?

Question... When I used **plot3d** to draw *one* parametric surface, I got *three* surfaces instead. What happened?

Answer... On some rare occasions, it can happen that your parametric surface is actually split into three pieces. However, most likely, you entered the command incorrectly. A common mistake is to type **{** *curly braces* **}** instead of **[** *square brackets* **]** for the coordinate functions. *Maple* will then interpret your input as a request to draw three *graphs* instead (see Chapter 15).

CHAPTER 20

Vector Fields

Drawing a Vector Field

The fieldplot Command

The **fieldplot** command (defined in the **plots** package) sketches vector fields in the plane. To see the vector field defined by $\vec{F}(x,y) = (F_1(x, y), F_2(x, y))$ for $x_0 \le x \le x_1$, $y_0 \le y \le y_1$, you enter:

```
with(plots);
    fieldplot([ F_1(x, y),  F_2(x, y) ], x = x_0..x_1, y = y_0..y_1);
```

For example, to see the vector field $(-y, x)$, for $-2 \le x \le 2$, $-2 \le y \le 2$, you type:

```
with(plots);
fieldplot([-y,x], x=-2..2, y=-2..2,
                  scaling=constrained);
```

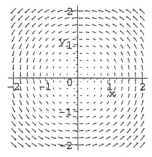

Strictly speaking, *Maple* does not draw the vector field $(-y, x)$ exactly. It has scaled the lengths of the vectors proportionally to produce a nice picture. Also, *Maple* draws a heavier arrow when the vector is longer.

Some Useful Options

Some options for the **fieldplot** command can help us change the style of the output picture.

Option	What It Does
`grid = [10,10]`	Draws $10 \times 10 = 100$ vectors. Default value is `[20, 20]`.
`arrows = thick`	Uses thicker arrows for vectors.
`color = blue`	The arrows will be drawn in blue (or whatever color you specify).

For example, here is a different picture of the vector field above:

122

```
fieldplot([-y,x], x=-2..2, y=-2..2, grid = [10,10],
              arrows = thick, color = grey);
```

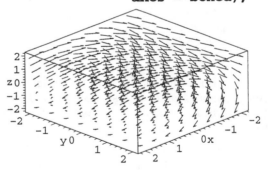

Vector Fields in Space

To draw a 3-D vector field (in space), you use the **fieldplot3d** command. It is defined in the **plots** package.

To see the vector field $\vec{F}(x, y, z) = (F_1(x, y, z), F_2(x, y, z), F_3(x, y, z))$, for the intervals $x_0 \le x \le x_1$, $y_0 \le y \le y_1$, and $z_0 \le z \le z_1$, you type:

```
with(plots):
fieldplot3d( [ F₁(x, y, z),  F₂(x, y, z),  F₃(x, y, z) ],
                x = x₀..x₁,  y = y₀..y₁,  z = z₀..z₁);
```

For example, to see the vector field $(-z, \ 1, \ x)$, use:

```
with(plots):
fieldplot3d( [-z,1,x], x=-2..2, y=-2..2, z=-2..2,
                axes = boxed);
```

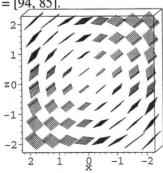

This looks like a mess. If you click on the picture and rotate the frame box, you can get a better view of the vector field by looking at it from the positive, *y*-axis, say with orientation = [94, 85].

Defining Vector Fields

The VectorCalculus package

Maple 8 introduces the **VectorCalculus** package with several very nice commands for working with vectors. To use these commands, you need to use the **VectorField** command, which specifies the coordinate system of the field.

```
with(VectorCalculus):
VectorField(<2*x+2*y,2*x,-3*z^2>,'cartesian'[x,y,z]);
```

$$(2x+2y)\bar{\mathbf{e}}_x + 2x\bar{\mathbf{e}}_y - 3z^2\bar{\mathbf{e}}_z$$

The symbols $\bar{\mathbf{e}}_x$, $\bar{\mathbf{e}}_y$, and $\bar{\mathbf{e}}_z$ that appear in the output above denote the standard basis vectors in space.

The **fieldplot** and **fieldplot3d** commands described in the previous section work with vector fields defined this way. Since you will do most of your work with vectors in a three-dimensional Cartesian coordinate system, it makes sense to set that coordinate system as the default.

```
SetCoordinates('cartesian'[x,y,z]):
VectorField(<2*x+2*y,2*x,-3*z^2>);
```

$$(2x+2y)\bar{\mathbf{e}}_x + 2x\bar{\mathbf{e}}_y - 3z^2\bar{\mathbf{e}}_z$$

> **Note.** The **VectorCalculus** package introduced in *Maple* 8 was written to be compatible with the new **LinearAlgebra** package. Users of earlier versions of *Maple* will need to use the old **linalg** package which has a number of tools for working with vectors and vector fields. We will describe the **linalg** package at the end of this chapter.

Other Coordinate Systems

We can also use a coordinate system other than the Cartesian coordinates in a computation. *Maple* understands the standard coordinate systems (Cartesian, polar, cylindrical, and spherical) and some esoteric ones as well (e.g. hyperbolic, rose, and toriodal).

To set the default system you use the **SetCoordinates** command with the name of the system and the variables used by that system. For example,

```
with(VectorCalculus);          #For Maple 8 or higher

SetCoordinates('cartesian'[x,y]):
SetCoordinates('cartesian'[x,y,z]):
SetCoordinates('polar'[r,theta]):
SetCoordinates('cylindrical'[r,theta, z]):
SetCoordinates('spherical'[rho,theta,phi]):
```

> **Note:** In the commands above, the order of the variables such as **[x, y, z]**, **[r, theta, z]**, and **[rho, theta, phi]** is important. If you list them in a different order, *Maple* will misinterpret your input.

We can also assign the coordinates used for a specified vector or vector field without changing the default coordinate system.

```
SetCoordinates('cartesian'[x,y]):     #Cartesian is the default
v1 := SetCoordinates( <3,4>, 'polar'[r,theta]);
```
$$3\overline{e}_r + 4\overline{e}_\theta$$

```
v2 := VectorField( <r*theta, theta>, 'polar'[r,theta]);
```
$$r\theta\overline{e}_r + \theta\overline{e}_\theta$$

Here, we are able to define a vector and a vector field in terms of $\overline{e}_r$ and $\overline{e}_\theta$ without changing the default Cartesian coordinate system.

Change of Coordinates

To convert a vector or vector field from one coordinate system to another, you use the **MapToBasis** command, naming the system you are converting to along with the variables used.

```
SetCoordinates('cartesian'[x,y]):
V1 := <3,4>;
V2 := MapToBasis(V1,'polar'[r,theta]):
simplify(%);
```
$$3\overline{e}_x + 4\overline{e}_y$$
$$5\overline{e}_r + \arctan\left(\tfrac{4}{3}\right)\overline{e}_\theta$$

```
SetCoordinates('cartesian'[x,y,z]):
V1 := VectorField(<2*x+z, y, z - x>);
V2 := MapToBasis(V1,'cylindrical'[r,theta,z]):
Simplify(%);
```
$$(2x+z)\overline{e}_x + y\overline{e}_y + (z-x)\overline{e}_z$$
$$(r\cos(\theta)^2 + \cos(\theta)z + r)\overline{e}_r - \sin\theta(r\cos\theta + z)\overline{e}_\theta + (z - r\cos\theta)\overline{e}_z$$

Gradient, Curl, and Divergence

Gradient Field

The gradient of a function is denoted by grad f. The gradient field of f always points in the direction of steepest increase of the value of f. In Cartesian coordinates it is defined to be the vector

$$\text{grad}\, f = f_x\overline{e}_x + f_y\overline{e}_y, \text{ or grad}\, f = f_x\overline{e}_x + f_y\overline{e}_y + f_z\overline{e}_z,$$

depending on whether f is a function of two or three variables, respectively. We find grad f with the **Gradient** function from the **VectorCalculus** package:

```
f := (x,y,z) -> x^2 + 2*x*y - z^3;
Gradient(f(x,y,z)) ;
```
$$(2x + 2y)\overline{e}_x + 2x\overline{e}_y - 3z^2\overline{e}_z$$

The gradient is often written in terms of the "del" or "nabla" operator, which is a vector of partial derivative operators.

$$\nabla = \left(\frac{\partial}{\partial x} \quad \frac{\partial}{\partial y} \quad \frac{\partial}{\partial z} \right)$$

We can also obtain the gradient by using either the **Del** or **Nabla** commands inside the **VectorCalculus** package.

```
Del(f(x,y,z));
```
$$(2x + 2y)\bar{\mathbf{e}}_x + 2x\bar{\mathbf{e}}_y - 3z^2\bar{\mathbf{e}}_z$$

Curl and Divergence

The curl and divergence of a vector field measure the rotation and spreading of particles whose motion is defined by the field. To calculate the curl and divergence of a vector field, use the **Curl** and **Divergence** commands, respectively.

```
Curl(VectorField(<x+y, y+z, sin(x+y)+z^2>));
```
$$(\cos(xy)x - 1)\bar{\mathbf{e}}_x - (\cos(xy)y)\bar{\mathbf{e}}_y - \bar{\mathbf{e}}_z$$

```
Divergence(VectorField(<x^2,y^2,z^2>));
```
$$2x + 2y + 2z$$

The curl and divergence can also be computed by taking the cross product and dot product, respectively, of the del operator with the vector field. Since **VectorCalculus** is already loaded, this can be done with the **CrossProduct** and **DotProduct**commands or with the **&x** and **.** operators, respectively.

We thus have three ways of computing these derivatives.

```
F := VectorField(<x+y, y+z, sin(x+y)+z^2>):
```

for curl, use any of	*for divergence, use any of*

`Curl(F);`
`CrossProduct(Del, F);`
`Del &x F;`
	`Divergence(F);`
	`DotProduct(Del, F);`
	`Del . F;`

Gradient, Curl, and Divergence in other coordinate systems

The **Gradient**, **Curl**, and **Divergence** commands can also be used in coordinate systems other than the Cartesian coordinate system. With **Gradient**, you simply specify the coordinate system and variable list.

Coordinate System	Maple *Command*
Gradient in the Cartesian system	`Gradient( x^2 + 2*x*y - z^3, [x,y,z]);` $(2x + 2y)\bar{\mathbf{e}}_x + 2x\bar{\mathbf{e}}_y - 3z^2\bar{\mathbf{e}}_z$

Gradient in the cylindrical system	`Gradient(r*cos(theta)+r^2*z,`
	`'cylindrical'[r,theta,z]);`
	$(\cos(\theta) + 2rz)\bar{\mathbf{e}}_r + (z - \sin(\theta))\bar{\mathbf{e}}_\theta - r^2\bar{\mathbf{e}}_z$

Gradient in the spherical system	`Gradient(2*rho^2*cos(theta)*sin(phi),`
	`'spherical'[rho, theta, phi]);`
	$(4\rho\cos(\theta)\sin(\phi))\bar{\mathbf{e}}_\rho + (-2\rho\sin(\theta)\sin(\phi))\bar{\mathbf{e}}_\theta$
	$-\left(2\frac{\rho\cos(\theta)\cos(\phi)}{\sin(\theta)}\right)\bar{\mathbf{e}}_\phi$

Since a vector field includes information of the coordinate system implicitly, it is not necessary to explicitly mention the coordinate system for **Curl** and **Divergence**.

Coordinate System	Maple Command
Divergence in the spherical system	```F := VectorField(<rho^2, rho*sin(phi), sin(theta)>,'spherical'[rho,theta,phi]): simplify(Divergence(F));``` $$\frac{4\,\rho\sin(\theta) + \sin(\phi)\cos(\theta)}{\sin(\theta)}$$
Curl in the cylindrical system	```G := VectorField(<theta, z, r>, 'cylindrical'[r,theta,z]): simplify(Curl(G));``` $$-\bar{e}_r - \bar{e}_\theta + \left(\frac{z-1}{r}\right)\bar{e}_\phi$$

Line and Flux Integrals

Integration of a Vector Field

Engineers are often interested in integrating a vector field $\vec{F}$ either along a curve $\vec{r}(t)$, for $a \le t \le b$, or through a surface $\vec{s}(u,v)$, for $u_0 \le u \le u_1$, $v_0 \le v \le v_1$. These are defined as:

- Line integral: $\displaystyle\int_a^b \vec{F}(\vec{r}(t)) \cdot \vec{r}'(t)\,dt$

- Flux integral: $\displaystyle\pm\int_{u_0}^{u_1}\int_{v_0}^{v_1} \vec{F}(\vec{s}(u,v)) \cdot \left(\frac{\partial \vec{s}}{\partial u} \times \frac{\partial \vec{s}}{\partial v}\right) dv\,du$ (The choice of the ± sign depends on how the normal vector for the surface is defined.)

These integrals are easily computed using the **LineInt** and **Flux** commands, inside the **VectorCalculus** package as we show in the following examples.

Line Integral

■ **Example.** Let $\vec{F}$ be the vector field $\vec{F}(x, y) = (x + y, -y)$ and $\vec{r}(t)$ the parametric curve $\vec{r}(t) = (1 - t, t^2)$ for $0 \le t \le 1$. The line integral of $\vec{F}$ over this curve is computed as:

```
with(VectorCalculus):
SetCoordinates('cartesian'[x,y]):
F := VectorField(<x+y, -y>):      # defines the vector field F.
```

Now, we have to define the curve $\vec{r}(t)$ using the **Path** command.

```
p := Path(<1-t,t^2>,t=0..1):
LineInt(F,p);
```
$$-\frac{4}{3}$$

We can see why the line integral turned out to be negative by checking how the vector field lines up with the curve.

```
with(plots):
field := fieldplot ( F, x=0..1, y=0..1,grid=[10,10]):
curve := plot( [1-t,t^2, t=0..1], thickness = 3):
```

```
display({field, curve}, scaling=constrained);
```

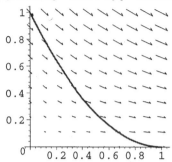

As you can see from the picture, if a particle moves along the curve from its starting point $\vec{r}(0) = (1,0)$ to its ending point $\vec{r}(1) = (0,1)$, it is moving against the force of the vector field. Thus, the line integral is negative.

Flux Integral

■ **Example.** Suppose the flux integral of $\vec{F}(x, y, z) = (x(1+z),\, y,\, 0)$ over the surface S is given by

$$-\int_{1}^{4}\int_{0}^{2\pi} \vec{F}(\vec{s}(u,v)) \cdot \left(\frac{\partial \vec{s}}{\partial u} \times \frac{\partial \vec{s}}{\partial v}\right) dv\, du$$

where S is parameterized by $\vec{s}(u,v) = (u\cos v,\, u\sin v,\, u)$. We can evaluate this integral as follows, noting that we use the **Surface** command to describe the parametric surface.

```
F := VectorField(-<x*(1+z),y,0>) :
s := (u, v) -> (u*cos(v), u*sin(v), u) :
Surf:=Surface(s(u,v), [u,v]=Rectangle(1..4,0..2*Pi):
Flux(F,Surf);
```

$$\frac{423\pi}{4}$$

Vector Calculus for *Maple* 7 and Earlier

linalg package

If you use an earlier version of *Maple* (*Maple* 7 or before), you must use the **linalg** package for vector calculus. In this case the vector field will be expressed in the form of a list. You must then use the **grad**, **curl**, **diverge** or other commands in the **linalg** package to perform an operation.

The following table gives examples of computing gradient, curl and divergence in the three most-used coordinate systems. But first, we load the **linalg** package:

```
with(linalg):
```

Example	Maple *Command for* Maple *7 and Earlier*
Gradient in the rectangular system	`grad( x^2 + 2*x*y - z^3, [x,y,z]);` $[2x+2y,\ 2x,\ -3z^2]$

Gradient in the cylindrical system	```grad(r*cos(theta)+r^2*z, [r,theta,z], coords = cylindrical);``` $[\cos(\theta)+2\,rz,\ -\sin(\theta),\ r^2]$
Divergence in the rectangular system	```diverge([x^2,y^2,z^2], [x,y,z]);``` $2\,x+2\,y+2\,z$
Curl in the rectangular system	```curl([x+y, y+z, sin(x*y)+z^2], [x,y,z]);``` $\left[\cos(x\,y)\,x-1,\ -\cos(x\,y)\,y,\ -1\right]$
Divergence in the spherical system	```diverge([rho^2, rho*sin(phi), sin(theta)], [rho,theta,phi], coords = spherical);``` $\dfrac{4\,\rho\sin(\theta)+\sin(\phi)\cos(\theta)}{\sin(\theta)}$

More Examples

The Perpendicular Property of the Gradient

■ **Example.** Consider the function $f(x,y)=xy+2x$:

```
f := (x,y) -> x*y +2*x;
```

The perpendicular property of the gradient vector states that $\operatorname{grad} f(a,b)$ is perpendicular to the level curve of f that goes through (a,b). To see this, we first draw several level curves:

```
pict1 := contourplot (f(x,y), x=-4..4, y=-4..4,
            contours = 20, color = black):
```

We can draw this gradient field with the **gradplot** command.

```
with(plots):
pict2 := gradplot(f(x,y), x=-4..4, y=-4..4,
            grid = [10,10]):
```

Now combine the contours and gradient vectors into one picture:

```
display({pict1, pict2}, scaling=constrained);
```

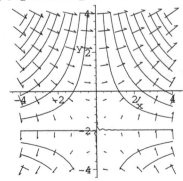

Do you agree that gradient vectors are perpendicular to the level curves everywhere?

Useful Tips

In many cases, you can draw a clearer (and faster) picture of vector fields by using a smaller number of gridlines, such as **grid = [10, 10]**.

The **VectorCalculus** package has commands for a number of other "derivatives" typically used in the subject. These include **Hessian**, **Jacobian**, and **Laplacian.** Check the appropriate help pages for more details. If a vector field is the gradient of a potential function, the **ScalarPotential** command can be used to recover the potential function. Similarly, if a vector field is the curl of a potential field, the **VectorPotential** command can be used to recover the potential field.

Instead of using a parametric **Path** to describe a given curve for the **LineInt** command, you may be able to define the curve more directly using one of the **Circle**, **Ellipse**, **Line**, **LineSegments**, or **Arc** commands. For example:

```
F := VectorField(<x+y, -y>, cartesian[x,y] ):
LineInt(F,Circle(<3,4>,2)); # a circle with center (3,4) radius 2.
LineInt(F, Line(<1, 1>, <2,5>));   # a line from (1,1) to (2,5).
```

Similarly, with the **Flux** command, you can use **Sphere** or **Box** to describe a surface without giving an explicit parameterization. For example:

```
F := VectorField(<x+y, -y, z>, cartesian[x,y,z] ):
```

To compute the flux of the vector field along the six boundary surfaces of the box $1 \le x \le 2, 3 \le y \le 4, 5 \le z \le 6$ with outward pointing normal, we use:

```
Flux(F, Box(1..2, 3..4, 5..6, outward));
```

To compute the flux along the sphere of center at (1,1,1) radius 3 with inward pointing normal, we use:

```
Flux(F, Sphere( <1,1,1>, 3, inward));
```

Troubleshooting Q & A

Question... I did not get a picture from **fieldplot** or **fieldplot3d**. What should I check?

Answer... There are several possible mistakes.

Input errors:

- Did you load the **plots** package? If you forgot, type the following:

```
with(plots);
```

- Did you mistakenly use **fieldplot** for a 3-D vector field or **fieldplot3d** for a 2-D vector field?
- Check that you have input the correct format and expression. For example, a common mistake is to enter the vector field $(x, x + y)$ as **(x, x+y)** instead of either **[x, x+y]** or **<x, x+y>**.

Hardware problems:

- Does the computer have enough memory to draw the picture (especially if you have already drawn a lot of graphics)? Restarting *Maple* may help.
- If your computer is slow, it can take a long time to show the output. Reducing the number of vectors by using the **grid** option can dramatically improve speed.

Question... When I used **Gradient**, **Curl**, or **Divergence**, (or **grad**, **curl**, or **diverge**) *Maple* returned my input unevaluated. What was my mistake?

Answer... You forgot to load the **VectorCalculus** (or **linalg**) package before you used these commands. Correct the problem by loading the package:

```
with(VectorCalculus);     #or
with(linalg);
```

Question... I got a wrong answer when I used **Gradient**, **Curl**, or **Divergence**, (or **grad**, **curl**, or **diverge**). What was my mistake?

Answer... There are several possibilities:

- Check that the input syntax is correct.
- You have to type the vectors **[x, y, z]**, or **[r, theta, z]**, or **[rho, theta, phi]**, either in setting coordinates or inside these commands, with the variables exactly in this order. If you list them in a different order, the computation of the gradient, curl, or divergence will be incorrect.
- The input function or vector field has to be expressed in terms of the variables of your chosen coordinates. For example, if you want to calculate the gradient of $x + y + z$ using cylindrical coordinates, you have to enter the function as **r*cos(theta)+r*sin(theta)+z**.

Question... When I used **Flux** or **SurfaceInt**, I got an error message about an unknown region of integration. What happened?

Answer... You probably misspelled the type of region you are integrating over. Check the syntax by looking at the help pages for **Flux**, **SurfaceInt**, and **int** in the **VectorCalculus** package.

CHAPTER 21

Basic Statistics

The **stats** (statistics) package in *Maple* contains a multitude of commands and procedures to analyze and describe data. Unlike the other packages that we have discussed so far, the **stats** package is organized into "subpackages" that can be called independently. In this chapter and the next, we will show you some basic statistics tools that you can use. Once you are familiar with the underlying concepts, the use of other advanced statistics tools is generally straightforward.

Graphical Presentation of Data

Scatter Diagrams

A set of numerical data can sometimes be presented graphically in a scatter diagram. Depending on the type of data, you can either use **pointplot** (defined in the **plots** package) or **scatterplot** (defined in the subpackage **statplots** of the **stats** package) to see the diagram.

```
with(plots);
pointplot( a list of ordered pairs );
```

```
with(stats[statplots]):
scatterplot( list of x-coordinates, list of y-coordinates );
```

> **Note:** You use the **pointplot** command if the data is a list of pairs. You use the **scatterplot** command if you have separate lists of *x*- and *y*-coordinates.

For example,

```
with(plots):
pointplot( [[1, 3], [2, 1], [6,2], [4, -1]]);
```

```
with(stats[statplots]):
scatterplot( [1,2,6,4], [3,1,2,-1]);
```

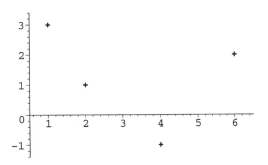

Numerical Measures of Data

Mean, Median, Standard Deviation, etc.

Standard statistical measures of data distribution are available in *Maple*. To use them, you first have to load the **describe** subpackage of the **stats** package:

```
with(stats[describe]);
```

> [*coefficientofvariation, count, countmissing, covariance, decile, geometricmean, harmonicmean, kurtosis, linearcorrelation, mean, meandeviation, median, mode, moment, percentile, quadraticmean, quantile, quartile, range, skewness, standarddeviation, sumdata, variance*]

■ **Example.** Let us consider the number of home runs for the 14 baseball teams in the National League for the 1997 season.

```
homeRuns := [174, 239, 142, 127, 136, 133, 174,
             172, 153, 129, 116, 152, 172, 144];
```

We can find the mean and median directly with the following commands.

```
mean(homeRuns);
```

$$\frac{309}{2}$$

```
median(homeRuns);
```

> 148

Other measures of central tendency such as the **harmonicmean**, **geometricmean**, and **mode** are also available and work in the same way.

The best-known measures of data variability also have their customary names:

```
evalf(variance(homeRuns));
```

> 891.5357143

```
evalf(standarddeviation(homeRuns));
```

> 29.85859531

```
evalf(meandeviation(homeRuns));
```

> 22.64285714

Maple's default on variance and standard deviation is for full populations. For example, the variance is calculated as $\frac{1}{N}\sum_{i=1}^{N}(X_i - \overline{X})^2$. If you want the sample

variance $\frac{1}{N-1}\sum_{i=1}^{N}(X_i - \overline{X})^2$, you have to add an index of 1 in the name of the

command:

evalf(variance[1](homeRuns));   # Sample variance.

960.1153846

evalf(standarddeviation[1](homeRuns));

30.98572872

Probability Distributions

The standard statistics packages contain definitions of common probability distributions. These include the following, along with their usual parameter specifications:

normald[μ , σ **]**	**uniform[** a, b **]**	**binomiald[** n, p **]**
studentst[ν **]**	**poisson[** λ **]**	**beta[** α, β **]**
exponential[α, a **]**	**discreteuniform[** a, b **]**	**chisquare[** n **]**

Cumulative Distribution and Probability Density Functions

You can work with the **cumulative distribution function**, the **inverse cumulative distribution function**, and the **probability density function** of each of these distributions using the **statevalf** command (defined in the **stats** package). The syntax is:

```
with(stats);
statevalf[ function type, distribution ]( arguments );
```

where the *function type* is listed in the following table:

Type of Distribution	function type	Meaning
Continuous	**cdf**	Cumulative distribution function
	icdf	Inverse cumulative distribution function
	pdf	Probability density function
Discrete	**dcdf**	Discrete cumulative distribution function
	idcdf	Inverse discrete cumulative distribution function
	pf	Probability density function

> **Note:** You have to use **[** *square brackets* **]** to surround the function type and distribution name in the **statevalf** command, but **(** *parentheses* **)** for the arguments of the command.

■ **Example.** Consider the normal distribution with mean 10 and deviation 3.

```
with(stats):
distribution1 := normald[10,3];
```

$$distribution1 := normald_{10,\,3}$$

To find the values of its cumulative distribution function and probability density function, say at $x = 15$, use:

```
statevalf[cdf,distribution1](15);
```

.9522096478

```
statevalf[pdf,distribution1](15);
```

.03315904623

We can see the graphs of the cumulative distribution, probability density, and inverse cumulative distribution functions, respectively, with the following. (Note that the domain of an inverse cumulative distribution function is $0 \le x \le 1$.)

```
plot(statevalf[cdf,distribution1](x), x=0..20);
```

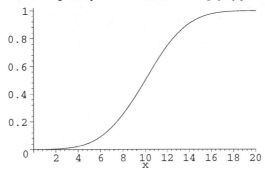

```
plot(statevalf[pdf,distribution1](x), x=0..20);
```

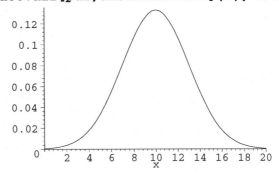

```
plot(statevalf[icdf,distribution1](x), x=0..1);
```

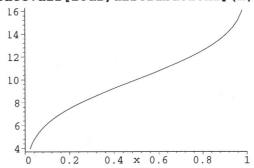

■ **Example.** We need to use **dcdf** and **pf** when working with discrete distributions such as the binomial distribution:

```
distribution2 := binomiald[15, 0.7]:
```

To find the value of its cumulative distribution function at 10:

```
statevalf[dcdf,distribution2](10);
```
>   .4845089408

We can also plot the "graph" of its probability density function:

```
pointplot([seq(
    [i, statevalf[pf,distribution2](i)],i=0..15)],
        title= "Binomial probs, n=15, p=.7",
        symbol = cross);
```

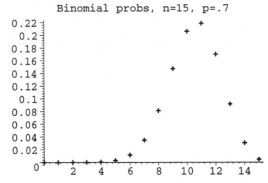

Binomial probs, n=15, p=.7

You can work with other distributions in a similar way. All you need to do is to include the proper parameter(s) when you reference the distributions.

More Examples

IQ Score

■ **Example**. It is generally believed that IQ scores are normally distributed with mean 100 and standard deviation 15. Let us theoretically estimate how many people in the world have an IQ higher than 180.

The probability that a person's IQ is between 0 and 180 is:

```
statevalf [cdf,normald[100,15]](180);
```
>   .9999999518

So the probability that a person's IQ is higher than 180 is:

```
1 - %;
```
> $.482 \ 10^{-7}$

There are almost six billion people in this world, so:

```
% * 6*10^9;
```
>   289.200000

Theoretically, there are fewer than 300 people in the world having IQs higher than 180. Isn't it odd, then, how often we meet people who claim to have such high IQs?

Generating a Random Sample from a Specified Distribution

■ **Example**. Using the command **random** defined in the **random** subpackage of the **stats** package, you can generate a random sample from specified distributions. (We will discuss more about random numbers in chapter 25.) For example, to form a list of 50 random values from the Poisson distribution with mean 5, we use:

```
with(random):
list1 := [random[poisson[5]](50)];
```

list1 := [4, 4, 4, 5, 5, 6, 1, 6, 5, 6, 3, 4, 7, 1, 2, 9, 7, 5, 6, 8, 9, 3, 3, 4, 5, 4, 3, 6, 7, 6, 4, 7, 6, 6, 4, 0, 2, 6, 3, 6, 3, 11, 5, 6, 5, 3, 7, 11, 2, 4]

The mean of this sample is:

```
mean(list1);
```

$$\frac{249}{50}$$ # This is pretty close to the theoretical value of 5.

■ **Example**. Let us consider **exponential[3,0]**. This is an Exponential distribution whose mean is $\frac{1}{3}$ and whose variance is $\frac{1}{9}$. Suppose we generate a sample of 1000 random observations from this distribution. We can see that the mean and variance of this random sample are very close to the theoretical values of $\frac{1}{3}$ and $\frac{1}{9}$, respectively:

```
list2 := [random[exponential[3,0]](1000)]:
mean(list2);
```

.3396637095

```
variance(list2);
```

.1138093963

Similarly for the uniform distribution **uniform[0,1]**, the mean should be very close to the center point 0.5.

```
list3 := [random[uniform[0,1]](1000)]:
mean(list3);
```

.5000200320

Importing Numerical Values from a File

You can easily import data values into *Maple* from another application (e.g., a spreadsheet or an e-mail message) and then use the statistical tools introduced in this chapter to study the data.

For example, suppose you have a data file named "TestScores.dat" where individual data values are written in text format, one to a line. You can read all the data values into a list named **dataValues** by using the **importdata** command as follows:

```
with(stats);
dataValues := [importdata("TestScores.dat")];
```

Now you can work with and analyze the data in the list **dataValues** by using commands from *Maple*'s statistics packages.

Useful Tips

☼ When you are working with statistical analysis, you will probably also want some descriptive statistics, the ability to evaluate a distribution, and some plotting tools. We recommend that you start by loading all the commands available in the statistics package with:

```
with(stats); with(describe); with(statevalf);
with(statplots);
```

You can load each individual statistics package separately. But if you load them all at the start of your *Maple* session, you are less likely to run into the usual troubles associated with unloaded commands.

Troubleshooting Q & A

Question... I tried to evaluate a cumulative probability function or probability density function at a point, but I got an error message such as "`... no subpackage function specified`" or "`... wrong number (or type) of parameters.`" What went wrong?

Answer... The statistics commands in *Maple* require special attention to the difference between brackets and braces. The command **statevalf** uses **[** *square brackets* **]** to enclose the function and the distribution. You use **[** *square brackets* **]** for the parameters of the distribution, but **(** *parentheses* **)** for the point at which you evaluate the distribution.

```
statevalf[cdf,normald[0,1]](3);
```

Question... Working on a desktop machine, when I tried to use **importdata** to load a file, I got an error message, "`... file I/O error.`" What went wrong?

Answer... The file may not exist, or it may not have the name you gave in the command. If it does exist and has the right name, check to see that is located in the folder or directory where *Maple* is looking for it.

The easiest way to verify where *Maple* looks for files is to use the **writedata** command to create a file with a distinctive name.

```
writedata("LookForMe",[1,2,3]);
```

Now, search for that filename on your disk. Start by looking for the file "LookForMe" in the folder with *Maple*.

CHAPTER 22

Fitting Curves to Data

When trying to fit a curve to a set of data points, there are two traditional approaches. You can either try to find a curve that hits the points exactly (**interpolation**) or find a curve of specified degree that gives a good approximation of the data (**regression**). *Maple* lets you use both approaches.

Regression

Using a Least Squares Fit

Given a list of data, you may want to find the line that "best" fits this data set. (Best fit is usually determined by the "least square method.") This process is called **linear regression**.

We can use the **LeastSquares** command from the **CurveFitting** package to carry out linear regression. (The **CurveFitting** package is not available for *Maple* 6 or earlier versions, we need to use the **leastsquare** command from the **stats[fit]** to get the job done. See the notes near the end of this chapter.)

```
with(CurveFitting):
LeastSquares( list of points , x);
```

or

```
LeastSquares( x-data , y-data, x);
```

For example, to find the line that best fits the points (1, 2), (3, 4), (5, 7), we have:

```
with(CurveFitting);
y = LeastSquares([ [1,2],[3,4],[5,7] ], x);
```

or

```
y = LeastSquares([1,3,5],[2,4,7], x);
```

$$y = \frac{5}{4} x + \frac{7}{12}$$

■ **Example.** Here are data from the UNESCO 1990 *Demographic Yearbook* which show male life expectancy and female life expectancy in Argentina, Austria, Afghanistan, Algeria, Angola, Bolivia, Brazil, Belgium, Bangladesh, and Botswana, respectively.

```
male := [65.5, 73.3, 41, 61.6, 42.9, 51, 62.3, 70,
              56.9,52.3]:
```

```
female := [72.7, 79.6, 42, 63.3, 46.1, 55.4, 67.6,
              76.8, 56,59.7]:
```

Now, let us find the line that best fits the data.

```
line1 := y = LeastSquares (male, female,x);
```

$$line1 := y = 1.128850251\, x - 3.192082489$$

To see how well the line fits the data, we first draw the points using **scatterplot** (see Chapter 21):

```
with(plots): with(stats[statplots]):
pict1 := scatterplot(male, female):
```

To plot the line, we need the right-hand side of the expression **line1** which is given by **rhs(line1)**.

```
pict2 := plot(rhs(line1), x=40..75):
display({pict1, pict2});
```

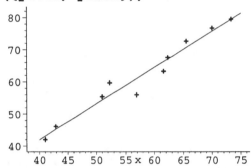

Under this model, say, if a certain country's male life expectancy is 65 years, then the female life expectancy is estimated to be:

```
subs(x=65, line1);
```

$$y = 70.18318383$$

Testing Linear Correlation

You use the command **linearcorrelation** (defined in the **describe** subpackage of the **stats** package) to find how well the best fitting line fits the data. Using our previous example:

```
with(stats[describe]):
linearcorrelation(male,female);
```

$$.9794660676$$

This value is very close to 1. Thus, a strong correlation exists between the life expectancy of males and females in the countries.

Fitting with Other Functions

The **LeastSquares** command can also work for other functions. Given functions $f_1(x), f_2(x), ..., f_n(x)$, if you want to fit a function of the form

$$a_1 f_1(x) + a_2 f_2(x) + ... + a_n f_n(x)$$

that best fits a given set of data, you will use one of these formats:

```
LeastSquares( x-data, y-data, x,
         curve = a_1 f_1(x) + a_2 f_2(x) + ... + a_n f_n(x));
```

$$\texttt{LeastSquares(} \textit{list of points}, \texttt{ x},$$
$$\texttt{curve} = a_1 f_1(x) + a_2 f_2(x) + \ldots + a_n f_n(x) \texttt{) ;}$$

For example, using the previous data:

```
LeastSquares(male,female,x,
            curve=a+b*x+c*x^2);          # Best quadratic fit.
```

$$5.744726622 + .8029638020 \ x + .002871857378 \ x^2$$

```
LeastSquares(male,female,x,
            curve=a+b*x+c*x^2+d*x^3); # Best cubic fit.
```

$$-122.1805302 + 7.812898348 \ x - .1222953133 \ x^2 + .0007297988519 \ x^3$$

```
LeastSquares(male,female,x,
            curve=a+b*exp(2*x)+c*exp(x));
```

$$57.79885201 - .1111944598 \ 10^{-60} \ e^{2x} + .7903142839 \ 10^{-29} \ e^x$$

The last answer suggests that it is not a good idea to use exponential functions to fit the given data in this case. Indeed, the coefficients of the exponential terms are virtually zero.

Using Weights in the Least Square Method

A drawback with the least squares method of curve fitting is that it is extremely sensitive to a single bad data point. For fitting linear equations, one alternative is to weight the data.

■ **Example.** Let us revisit the UNESCO data from our earlier example but add in an extra point at (5, 85).

```
UNESCO1 := [seq([male[i],female[i]],i=1..10)]:
                                      # Original data as a list of points.
UNESCO2 := [op(UNESCO1), [5,85]]:    # With extra point.
reg1 := LeastSquares(UNESCO1,x);     # The original data.
```

$$reg1 := 1.128850251 \ x - 3.192082489$$

```
reg2 := LeastSquares(UNESCO2,x);     # With the bad point.
```

$$reg2 := .02684747952 \ x + 62.59819422$$

Notice how different these results are. One bad point has affected the result dramatically.

One remedy is to give less weight to points considered to be suspect. The weights must be positive integers and the list of weights should be the same length as the list of points. (In the command below, we use **nops** to determine the number of elements in the original list.)

```
reg3 := LeastSquares(UNESCO2,x,
        weight=[seq(20,i=1..nops(UNESCO1)),1]);
```

Here the "good points" are each weighted with 20, and the "bad point" is weighted with 1 only.

$$reg3 := .94894074134 \ x + 7.548631364$$

Adding one bad point with less weight than the others gives a line that resembles the line obtained without the point.

Interpolation

The interp Command

Given n points (x_1, y_1), (x_2, y_2), ..., (x_n, y_n), you can find a polynomial of degree $n-1$ that perfectly matches all these values. This can be done in *Maple* using the command **PolynomialInterpolation** from the **CurveFitting** package in the syntax:

```
with(CurveFitting):
PolynomialInterpolation( x-data, y-data, x);
PolynomialInterpolation( list of points, x);
```

(For *Maple* 6 or earlier versions, you need to use instead the **interp** command. Please see the discussion near the end of the chapter.)

For example, to find a polynomial $g(x)$ that passes through all of the following points, use:

```
points:= [[0,0], [1,16], [2,-10], [3,28],
          [4,-30], [5,-12], [6,-12]]:
with(CurveFitting);
g := PolynomialInterpolation(points, x);
```

$$g := -\frac{299}{180}x^6 + \frac{299}{10}x^5 - \frac{1819}{9}x^4 + \frac{1897}{3}x^3 - \frac{162041}{180}x^2 + \frac{13733}{30}x$$

You can see that the graph of $g(x)$ passes through all the given points with:

```
with(plots):
pict1 := pointplot(points, symbol=cross):
pict2 := plot(g, x=0..6):
display({pict1,pict2});
```

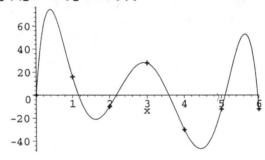

The Spline Command

Given n points (x_1, y_1), (x_2, y_2), ..., (x_n, y_n), another approach to curve fitting is to look for a curve that is defined "piecewise" by polynomials of a given degree d. This is done with the **Spline** command.

```
with(CurveFitting):
Spline( list of points, x, degree = d);
```

We can compare the outputs of **Spline** and **PolynomialInterpolation** on the same set of points below.

```
data:= [[0,0],[1,10],[2,4],[3,10],[4,1],[5,10]]:
with(CurveFitting):
c3 := Spline(data, x, degree = 3):
```

```
c4 := PolynomialInterpolation(data, x):
pict1 := plot([c3,c4],x=0..6,y=-4..14,
              legend=["spline","poly"]):

with(plots):
pict2 := pointplot(data,
              symbol=cross, legend="data points",
symbol = 24):
display({pict1, pict2});
```

Curve Fitting in *Maple* 6 and Earlier

Maple 6 (and earlier versions) does not have the **CurveFitting** package. Other commands must be used for regression and interpolation.

LeastSquares vs. leastsquare

In *Maple* 6 (and earlier versions), you must use the **leastsquare** command from the **stats[fit]** package to carry out linear regression. It is used in the form:

```
with(stats[fit]):
leastsquare[[x,y], y = a*x+b ]([ x-data, y-data ]);
```

For example,

```
with(stats[fit]);
leastsquare[[x,y], y=a*x+b ]([ [1,3,5], [2,4,7] ]);
```

$$y = \frac{5}{4} x + \frac{7}{12}$$

Similarly, the **leastsquare** command can also work for other functions. Using the UNESCO data that we used at the beginning of the chapter:

```
leastsquare[[x,y],y=a+b*x+c*x^2]
              ([male,female]);      # Best quadratic fit.
```

$$y = 5.744726622 + .8029638020\ x + .002871857378\ x^2$$

```
leastsquare[[x,y],y=a+b*x+c*x^2+d*x^3]
              ([male,female]);      # Best cubic fit.
```

$$y = -122.1805302 + 7.812898348x - .1222953133x^2 + .0007297988519x^3$$

```
leastsquare[[x,y],y=a+b*exp(2*x)+c*exp(x),{a,b,c}]
    ([male,female]);
```

$$y = 57.79885201 - .1111944598 \ 10^{-60} \ e^{2x} + .7903142839 \ 10^{-29} \ e^x$$

Unfortunately, **leastsquare** does not accept input of data as a list of points, and you cannot weight the data to avoid bad data points. These are reasons why you'll prefer to use **LeastSquares** if you use *Maple* 7 or higher.

Polynomial Interpolation vs. interp

In *Maple* 6 (and earlier versions), you must use the **interp** command for polynomial interpolation. It has the syntax:

```
interp( x-data, y-data, x );
```

For example, to find a polynomial $g(x)$ that passes through all of the points, use:

```
xdata := [0, 1, 2, 3, 4, 5, 6]:
ydata := [0, 16, -10, 28, -30, -12, -12]:
g := interp(xdata, ydata,x);
```

$$g := -\frac{299}{180}x^6 + \frac{299}{10}x^5 - \frac{1819}{9}x^4 + \frac{1897}{3}x^3 - \frac{162041}{180}x^2 + \frac{13733}{30}x$$

The **interp** command will not accept the input of data as a list of points. This makes it less convenient to use than the **PolynomialInterpolation** command.

More Examples

Fitting vs. Interpolating-Polynomial

You may wonder about the difference between the methods of least squares fitting and interpolation. A least square fit allows us to find a function of a specified form that best *approximates* the given set of data. On the other hand, interpolation will find a function that *matches the data exactly*. A least square fit is normally used when we do not believe the independent variable entirely predicts the dependent variable.

■ **Example.** Consider the UNESCO data of the first example of this chapter. If we had used interpolation rather than finding a least squares fit for this data, we would have found the polynomial:

```
Digits := 12;     # We need to increase the number of digits used in
                  # the floating point computation, because the answer
                  # behaves so wildly, as you will see.

with(CurveFitting):
PolynomialInterpolation(male,female,x);
```

$$-.397572075218 \ 10^{-7}x^9 + .203073369106 \ 10^{-4}x^8 - .458829289271 \ 10^{-2}x^7$$
$$+.601845366123 \ x^6 + 50.5044284563 \ x^5 + 2811.63405653 \ x^4 - 103837.722471x^3$$
$$+2453065.00239 \ x^2 - 33636458.2451 \ x + 203962232.163$$

Let us compare graphically the results obtained by the two methods:

```
with(plots): with(stats[statplots]):

pict1 := scatterplot(male, female):
pict2 := plot(LeastSquares(male,female,x),
       x=40..75, color=green):
pict3 := plot(PolynomialInterpolation (male,
      female, x), x=40..75,y=30..80, color=grey):
display({pict1, pict2, pict3});
```

You can see in this case that the line obtained by least squares fitting is likely to be better for prediction.

A Temperature Model

■ **Example.** The following monthly data in the form [*month, temperature in* $F°$] show the average monthly high temperatures in Boston, Massachusetts (1 represents the month of January, 2 for February, and so on).

```
bostonHigh := [[1, 36.4], [2, 37.7], [3, 45.0],
               [4, 56.6], [5, 67.0], [6, 76.6],
               [7, 81.8], [8, 79.8], [9, 72.3],
               [10, 62.5],[11,51.6], [12,40.3],
               [13, 36.4],[14, 37.7],[15, 45.0]]:
pointplot(bostonHigh, symbol=cross);
```

Since temperature is cyclic, we expect that it can be described nicely by a cosine function in the form

Temperature $= a_1 + a_2 \cos(a_3(x - a_4))$, where x is the month.

The temperature's period is 12 months, so we should choose $a_3 = \frac{2\pi}{12} = \frac{\pi}{6}$. Also, since we expect July to be the hottest month of the year, the function must have its largest value when $x = 7$. This means that we should choose $a_4 = 7$.

Thus, we need to find the function of the form $a_1 + a_2 \cos(\frac{\pi}{6}(x - 7))$ that best fits the temperature data.

```
bestfit := LeastSquares(bostonHigh,x,
curve=a+b*cos(Pi/6*(x-7)), params={a,b});
```

$$bestfit := y = 58.7015403930 + 22.7483212182\cos\left(\frac{1}{6}\,\pi\,(x-7)\right)$$

> **Note:** We need to include **params={a,b}** to specify the parameters, otherwise *Maple* will think **Pi** is an indeterminate and gives an error message that the equation is not linear. If we replace **Pi** with **evalf(Pi)**, then we do not need to include **params={a,b}**.

```
plot1 := pointplot(bostonHigh, symbol=cross):
plot2 := plot(bestfit,x=0..16):
display({plot1, plot2});
```

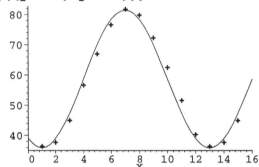

The temperature data fit nicely in this model. (The result can be even more impressive if we choose $a_4 = 7.25$ instead. Try it yourself!)

Useful Tips

It is often useful to plot data points along with the curve you fit to the data. Two commands are useful in plotting the data points. If the data is as a list of points, use the **pointplot** command from the **plots** package. If the *x*-data and *y*-data are in separate lists, use the **scatterplot** command from the **stats[statplots]** package.

In order to diminish the effect of a single bad data point during the least square method, one can instead use the **leastmediansquare** command in the **stats[fit]** package. It has the same syntax as the **leastsquare** command.

If you are planning to do a lot of statistics, we recommend that you also read Chapters 24, 25, and 26, so you can learn more about lists, random numbers, simulation, and spreadsheets. For example, you will learn to use the **zip** command to create ordered pairs from data lists and use the **map2** command to create data list from ordered pairs.

Troubleshooting Q & A

Question... I used the **LeastSquares** command, but got error messages such as "`... 3rd argument must have type name,`" or "`... data points not in recognizable format.`" What went wrong?

Answer... You probably have a problem with brackets. Either you forgot to put brackets around a list of points (giving too many arguments), or you put brackets around your lists of *x-data*, *y-data* (giving too few arguments).

Question... I used the **LeastSquares** command, but got error messages such as "`... number of values in first and second arguments must match.`" What went wrong?

Answer... Check that your lists of *x-data* and *y-data* have the same number of entries.

Question... When I tried to graph the points of a data set along with the polynomial obtained from using **PolynomialInterpolation**, the curve did not go through the points. What went wrong?

Answer... Inaccuracies produced by **PolynomialInterpolation** are often due to roundoff error in the calculation of the coefficients of the polynomial. You may need to increase the number of digits used in the floating point computation. For example, try:

```
Digits := 12;
```

CHAPTER 23

Animation

Getting Started

The animate Command

It is fun and easy to do animation in *Maple*. For 2-D animations, you can use the **animate** command in the following form (it is defined in the **plots** package):

```
with(plots);
animate( a function of t and x, x = x_0..x_1, t = t_0..t_1 );
```

The *function* inside **animate** can be any object that will work in the **plot** command that we discussed in earlier chapters. For example, type

```
with(plots):
animate( sin(t*x), x = 0..2*Pi, t = 1..10);
```

Now, click on the picture, a control strip will show up near the top of the worksheet in the context bar. It has the control buttons as that of a VCR. Click the "play" button ▶ on the control strip to see the animation.

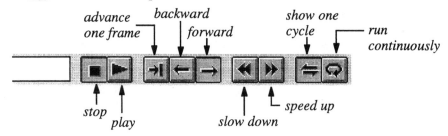

How does the Animation work?

In the example above, when you type:

```
animate( sin(t*x), x = 0..2*Pi, t = 1..10);
```

Maple first produces frames of the animation for numerous values of *t* between 1 and 10. These frames are approximately the same as the output of the commands:

```
plot(sin(1*x),x = 0..2*Pi);      # frame 1.
plot(sin(1.6*x),x = 0..2*Pi);    # frame 2.
# etc.
```

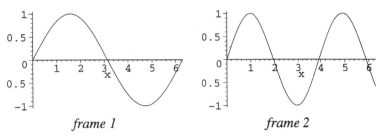

frame 1   *frame 2*

After *Maple* finishes creating all the pictures (sixteen by default), it then displays them rapidly enough so that your eye sees a continuous motion.

The **animate** command works the same way as the **plot** command. The following demonstrate some variations of the **animate** command.

```
animate({t*x, -t*x, t*x^2, -t*x^2, t*x^3, -t*x^3},
        x=-1..1, t=-4..4);
```

(Here, each frame consists of the graphs of tx, $-tx$, tx^2, $-tx^2$, tx^3 and $-tx^3$.)

```
animate(
    {[t*cos(x), t*sin(x), x=0..2*Pi],
     [(1-t)*cos(x)+2, (1-t)*sin(x), x=0..2*Pi]},
    t=0..1, scaling = constrained);
```

(In this case, each frame consists of the two parametric circles $(t\cos x, t\sin x)$ and $((1-t)\cos x + 2, (1-t)\sin x)$.)

Animating a 3-D Picture

You can make 3-D animations in *Maple* as well. You simply use the **animate3d** command just as you used the **animate** command. For example,

```
with(plots);
animate3d(sin(t*x)+cos(t*y),x=0..Pi, y=0..Pi,t=1..5,
    axes=boxed);
```

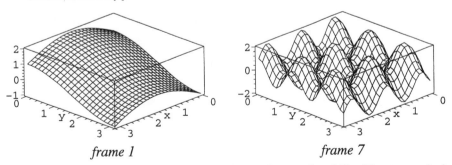

frame 1   *frame 7*

We show only two frames of the animation above (frames 1 and 7). If you watch the animation, you will see "waves" move through the surface.

Advanced Examples

The display Command

The examples we have looked at so far have been simple animations in which each frame could be plotted with a single **plot** command.

When we want to make more complicated animations, we must make double use of the **display** command -- once to make the frame, then a second time to provide the animation.

Strategy

The key to writing sophisticated animations is to have a firm grasp of what each frame looks like, as well as how the frames vary according to a parameter. Following this concept, you can more easily create a complicated animation.

In each of the following examples, we'll define **oneFrame**, a "frame function" of t, that will generate the graphic elements of just one frame. Because each frame itself will be a combination of several simple graphics, we'll use the **display** command within the definition of **oneFrame**:

```
oneFrame := t->display([ graphics command(s) depending on t ]);
```

Then we will use the **display** command a second time with a sequence of frames produced by **oneFrame**. This time we set the **insequence** option to **true** to provide animation.

```
with(plots);
display([ a sequence of frames ], insequence=true);
```

Moving a Point Along a Curve

■ **Example**. Let us show you how to write the animation of a point moving along the sine graph, $y = \sin x$. The coordinate of the point at any time t can be thought as $P(t) = (t, \sin(t))$ for $0 \le t \le 2\pi$.

In each frame of the animation, we will plot the position of the point at time t against a background of the sine curve. The sine curve can be sketched with a **plot** command, and the point can be drawn using a **pointplot** command. We combine them using **display**.

```
oneFrame := t -> display([ plot(sin(x),x=0..2*Pi),
        pointplot([[t, sin(t)]], symbol=circle)]):
```

For example, you can see the frame at $t = 0.5$ with:

```
oneFrame(0.5);
```

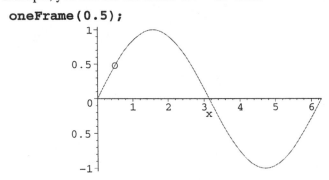

The frames are then made into a sequence that is animated with a second use of the **display** command. To animate over the interval $0 \le t \le 2\pi$ with 15 frames we use:

```
display([seq(oneFrame(i*2*Pi/14),i=0..14)],
                            insequence=true);
```

Rolling a Ball Along the Ground

■ **Example**. A circle of radius one "rests" on top of the x-axis at the origin, as pictured below. A point P, which is marked at the "top" of the circle, is currently touching the y-axis at the point $(0, 2)$.

As the circle begins to "roll" to the right, the point P will rotate downward and eventually hit the x-axis after the circle has rolled a distance of π (because P was originally halfway around a circle of circumference 2π). Thereafter, P will rotate upward again after the circle "rolls over" it, reaching a high point again when the circle has rolled a distance of 2π.

The following suggests the motion of P, where a dot shows the location of P:

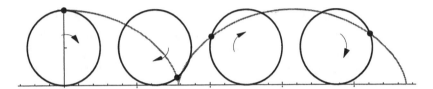

In fact, the path followed by P as the circle rolls is known to be a cycloid given parametrically by $(u + \sin(u), 1 + \cos(u))$, for $0 \le u$.

We will let the parameter t in this animation be the distance the ball has rolled. Each frame of the animation should have three elements:

- A fixed portion of the cycloid to act as a background. It can be given by $(u + \sin(u), 1 + \cos(u))$, for $0 \le u \le 4\pi$.

- The new position of the unit circle in this frame. Since the circle has moved a distance t, its center is at $(t, 1)$ and its equation is thus $(t + \cos(u), 1 + \sin(u))$.

- The corresponding position of the point P. After the ball has traveled a distance of t, P will have rotated a distance of t clockwise around the circle. P's location relative to the center of the circle is thus given by $(\sin(t), \cos(t))$. Its position in the frame will then be $(t + \sin(t), 1 + \cos(t))$.

After some experimentation with **view** to adjust the range of display for the x- and y- coordinates, our frame function is as follows.

```
with(plots):
oneFrame := t -> display([
    plot({[u+sin(u), 1+cos(u), u=0..4*Pi],
            [t+sin(u), 1+cos(u), u=0..2*Pi]}),
    pointplot([[t+sin(t), 1+cos(t)]],
            symbol=circle)],
            view = [-1..10,-.1..2.1],
            scaling=constrained):
```

The completed animation, which rolls P over two arcs of the cycloid, is given with:

```
display([seq(oneFrame(i*4*Pi/16),i=0..16)],
            insequence=true, scaling=constrained);
```

The Oscillating Spring

■ **Example**. A metal plate is attached to a spring of height h_0. (The picture to the right shows such a spring at a height of 5 units above the xy-plane.) The plate is then displaced upward a distance of c units and released. The plate (and spring) will begin a vertical, damped oscillation.

If we ignore the effect of gravity and the weight of the plate, then the height of the plate at any time t after release will be

$$z = h_0 + c\,e^{-bt}\cos(kt),$$

where b and k are constants that depend on the spring.

We'll model the motion of the plate using animation in *Maple*. The graphic elements of each frame are as follows.

- The height of the "plate" is $z = h_0 + c\,e^{-bt}\cos(kt)$. It can be seen with **plot3d**. We'll sketch it only over the rectangle $-1 \leq x \leq 1$, $-1 \leq y \leq 1$.

- We'll think of the "spring" as the helix $(\cos(u), \sin(u), \frac{u}{8\pi}(h_0 + ce^{-bt}\cos(kt))$, for $0 \leq u \leq 8\pi$, and sketch it using **spacecurve**.

Here is the frame function:

```
oneFrame := t -> display([
    spacecurve([cos(u), sin(u),
                u*(h0+c*exp(-b*t)*cos(k*t))/(8*Pi)],
                u=0..8*Pi, axes=boxed),
    plot3d(h0+c*exp(-b*t)*cos(k*t),
                x=-1..1, y=-1..1)]):
```

Before you can see an animation, you must choose numerical values for the various constants h_0, c, b, and k we have introduced. For example:

```
h0 := 5;       # The plate is at height 5 at rest.
c  := 2;       # It is displaced 2 units upward.
b  := 0.05;    # This coefficient controls damping.
k  := 1;       # This coefficient controls the speed of oscillation.
```

Now you can do the animation, say, over the time interval $0 \leq t \leq 8$:

```
display([seq(oneFrame(i*8/16),i=0..16)],
        insequence=true);
```

Try repeating the animation for different numerical values of the constants and see how each one affects the movement of the plate. For example, increasing the value of b will cause the oscillation to die out more quickly.

Useful Tips

The **plot** command has many options. Some, like **style**, apply to the object being graphed. Others, like **titlefont**, apply to background settings. However, the **display** command cannot reconcile conflicting background settings from the various plot structures that make up an animation. It simply chooses some and ignores others. Adding *identical* options like **axes**, **title**, and **tickmarks** to each plot structure will help you control the output of an animation.

Troubleshooting Q & A

Question... How do I decide whether to use **animate** or **display**?

Answer... Since **animate** is simpler and easier to use than the double use of the **display** command, use **animate** and **animate3d** whenever you can.

You can use **animate** and **animate3d** whenever you can graph a single frame with the **plot** or **plot3d** command. This will be the case when you simply want to animate the graph of a function, or even several functions, with a time parameter.

If you need other graphing commands, like **pointplot**, **implicitplot**, or **spacecurve** to graph a single frame, you should use the **display** command. The **display** command is always a better choice when you want to have explicit control over the frames in the animation.

Question... I tried using the **display** command, but I got pages of numbers instead of a picture. What happened?

Answer... When the **display** command fails, it fails spectacularly. It shows the input it has received, which is a series of plot objects. Each plot object is a long sequence of numbers that tells *Maple* what it is trying to plot, including all the points on the graph and all the tickmarks on the axes.

The animations we have been looking at are compound plot structures. If the animation doesn't work, check the pieces.

Use a command like **display([oneFrame(1)]);** to test if individual frames graph properly. If the frames graph properly, the difficulty is with the syntax of the **display** command that does the animation. The most common errors are misspelling a keyword and forgetting to use [*square brackets*] to enclose the sequence of frames.

If the problem is with an individual frame, try plotting the elements separately for a variety of values of the parameter *t*.

Question... *Maple* ran out of memory before completing the animation. What should I do?

Answer... Animations use lots of memory (especially in 3-D).

Try increasing the memory allocation for *Maple* . Closing unneeded windows and quitting other applications before restarting may help too.

Also, try your animation using a small number of frames. Gradually increase the number of frames until you get a satisfactory result.

CHAPTER 24
More About Lists

In this chapter we will show you more commands that you can use when you work with lists. Many of them will be helpful in statistics or *Maple* programming.

Basic Lists

Setting Up and Reading Lists

We have been using lists throughout this text. (See chapter 6.) At a basic level, a list is an ordered sequence enclosed in square brackets. The entries of a list can be any valid *Maple* object.

For example, consider the list,

```
myList := ["red", 2, 3, 3.153, x^2=5, sin(z),
              Pi, "history", Fred];
```

$$myList := [\text{ "red", } 2, 3, 3.153, \ x^2 = 5, \ \sin(z), \ \pi, \text{ "history", } Fred]$$

Elements of **myList** include **"strings of words"**, a name, integers, real numbers, an equation, and a function. Each element can be recalled by **myList[1]**, **myList[2]**, and so on.

```
myList[5];
```

$$x^2 = 5$$

The syntax **myList[-n]** denotes the element of **myList** that is *n* entries from the end of the list.

```
myList[-3];
```

$$\pi$$

We can use **myList[a..b]** to specify a range of entries of **myList**.

```
myList[3..6];                # Shows the third to sixth elements.
```

$$[3, 3.153, \ x^2 = 5, \sin(z)]$$

Note that the result of **myList[a..b]** is given in a list structure. If you want to see the elements without a list, then you use the **op** command:

```
op(myList[3..6]);
```

$$3, 3.153, \ x^2 = 5, \sin(z)$$

If an element of a list is a list itself, we can retrieve its entries in the same manner, using [*square bracket*] syntax. For example,

```
newList := [[1, 5, 7], [-6, 5, 7,7], 4, 6, 8]:
```

```
newList[2];                # The second element of newList is a list itself.
```

$$[-6, 5, 7, 7]$$

newList[2][1..3];   # We can also find the elements of an element.

$$[-6, 5, 7]$$

Modifying the Elements of a List

To change an entry in a list, you can use a normal assignment statement. Recall the list that we used earlier.

myList;

$$[\text{"red"}, 2, 3, 3.153, \ x^2 = 5, \ \sin(z), \ \pi, \ \text{"history"}, \ Fred \ ]$$

We want to replace the first element of this list, say, with 27:

myList[1] := 27;

$$myList_1 := 27$$

Now you can see that the first element of **myList** has been replaced.

myList;

$$[\ 27, 2, 3, 3.153, \ x^2 = 5, \ \sin(z), \ \pi, \ \text{"history"}, \ Fred \ ]$$

We can also modify a list directly by redefining its elements:

**myList := [op(myList[1..3]), "good", "bad", "ugly",
                op(myList[5..7]), 100, 101];**

$$myList := [27, 2, 3, \text{"good"}, \text{"bad"}, \text{"ugly"}, \ x^2 = 5, \ \sin(z), \ \pi, \ 100, \ 101]$$

The new **myList** consists of the first three elements of the original list, then the new elements **"good"**, **"bad"**, and **"ugly"**, followed by the fifth to seventh elements of the original list, and then the numbers 100 and 101.

As an example, it might be of interest to construct a list of the first 1000 prime integers, using the **ithprime** command. We find these with:

lotsOfPrimes := [seq(ithprime(i), i=1..1000)]:

We can see the first three and last three elements of this list with:

**[op(lotsOfPrimes[1..3]),
op(lotsOfPrimes[998..1000])];**

$$[2, 3, 5, 7901, 7907, 7919]$$

You were probably already familiar with the first few primes, but not many people realize that 7919 is the 1000th prime.

Useful List Commands

The map Command

To evaluate a function at each element of a list, you use the **map** command. It has the form:

map(*function name* **,** *the list* **);**

Let us look at an easy example:

```
list1 := [ seq(2^n, n=1..8)];
```
$$list1 := [2, 4, 8, 16, 32, 64, 128, 256]$$

```
f1 := x -> x+1 :                    # We want to add 1 to each element.
map(f1, list1);
```
$$[3, 5, 9, 17, 33, 65, 129, 257]$$

> **Note:** Inside the **map** command, we type the *name* of the function only. In the example above, we use **f1** and not **f1(x)**.

Now, we'll try a more interesting example:

```
list2 := [ [`John`, 75, 62], [`David`, 62, 81],
           [`Mary`, 75, 91], [`Jane`,  31, 50],
           [`Steve`,21, 31]]:
```

This shows a list of five students and their scores in two exams. We will calculate the average score of each student and list it with their name:

```
f2 := x -> [x[1], (x[2]+x[3])*0.5]:
```

(Note that **x[1]** is the name of the student, while **x[2]** and **x[3]** are the exam scores.)

```
list2a := map(f2,list2);
```
$$list2a := [[John, 68.5], [David, 71.5], [Mary, 83.0], [Jane, 40.5], [Steve, 26.0]]$$

You can also use **map** to evaluate a multi-variable function with the first argument coming from the elements of a list.

> **map(** *function name* , *the list* , *argument 2* , *argument 3* , *etc.* **);**

For example,

```
f1a := (x,y,z) -> x^y+z;          # a function of three variables
map(f1a, [seq(i,i=1..5)],2,1);    # for each of the numbers in a
                                   # list, square it (^2) and add 1
```
$$[2, 5, 10, 17, 26]$$

The map2 Command

Similarly, there is a **map2** command where a list is used for the second argument of a multi-variable function.

> **map2(** *function name* , *argument 1* , *the list* , *argument 3* , *etc.* **);**

```
map2(f1a,2,[seq(i,i=1..5)],1);
```
$$[3, 5, 9, 17, 33]$$

If you are given a list of data points in coordinates, you can easily use the **map2** and **op** commands together to extract the values of all the *x*- or *y*-coordinates. For example:

```
pairs := [ [1,2], [3, 4], [5, 6], [7, 8], [9, 10]]:
map2(op,1, pairs);                    #extract the x-values
```
$$[1, 3, 5, 7, 9]$$

```
map2(op,2, pairs);                    #extract the y-values
```
$$[2, 4, 6, 8, 10]$$

The **map** (or **map2**) command provides an easy way to implement "data transformation" or "data massaging" if you work with statistical data. See the US population model in the More Examples section.

The zip Command

The **map** and **map2** commands allow only one of the arguments of a multi-variable function to come from a list. The **zip** command can be used if both of the arguments of a two-variable function are lists.

zip(*function name* **,** *first list* **,** *second list* **);**

For example,

```
f3 := (x, y) -> x*y:
a := [1,2,3,4,5]:
b := [6,7,8,9,10]:
zip(f3, a, b);
```

$$[6, 14, 24, 36, 50]$$

If you are given data lists of *x*-coordinate and *y*-coordinate separately, use the **zip** command to combine them to form the corresponding ordered pairs.

```
xValue := [1, 3, 5, 7 ,9] :
yValue := [2, 4 ,6 ,8, 10]:
f := (x, y) -> [x, y] :

zip(f, xValue, yValue);
```

$$[[1, 2], [3, 4], [5, 6], [7, 8], [9, 10]]$$

Sorting Lists

The **sort** command will arrange the elements of a list in increasing order. It has the syntax:

sort(*list* **);**

For example,

```
list3 := [81, 70, 97, 63, 76, 38, 85, 68, 21]:
sort(list3);
```

$$[21, 38, 63, 68, 70, 76, 81, 85, 97]$$

You can also specify any "Boolean function" on two variables as the sorting criterion. It has the syntax:

sort(*list* **,** *Boolean function* **);**

For example, you can sort the **list3** above in descending order.

```
f3 := (a,b) ->          # This Boolean function will return true if the
        evalb(a > b):   # first element is larger than the second element.
sort(list3, f3);
```

$$[97, 85, 81, 76, 70, 68, 63, 38, 21]$$

This flexibility lets us sort lists whose elements are lists themselves. Consider the following example that sorts our list of students according to their second exam scores:

```
list2 := [ [`John`, 75, 62], [`David`, 62, 81],
           [`Mary`, 75, 91], [`Jane`, 31, 50],
           [`Steve`,21, 31]]:
```

```
f4 := (x,y) -> evalb(x[3] < y[3]):
```

(Note that in this function, **x** and **y** refer to a list [*name, exam1, exam2*]; thus **x[3]** and **y[3]** are the third elements, *exam2*.)

```
sort(list2, f4);
```

[[*Steve*, 21, 31], [*Jane*, 31, 50], [*John*, 75, 62], [*David*, 62, 81], [*Mary*, 75, 91]]

Counting the Elements

The command **stats[transform, tally[count]]** shows how many times each item appears in a list. It has the syntax:

```
stats[transform, tally[count]]( your list );
```

For example,

```
list1 := [ 3, 3, 5, 6, 5, 9, 10, 5, 5, 1, 5, 1, 5];
```

list1 := [3, 3, 5, 6, 5, 9, 10, 5, 5, 1, 5, 1, 5]

```
stats[transform, tally[count]](list1);
```

[6, 9, 10, Weight(1, 2), Weight(3, 2), Weight(5, 6)]

This means that 6, 9, and 10 appear once, 1 appears twice, 3 appears twice, and 5 appears six times.

Other List Commands

Many other commands for lists will be useful, especially for programming in *Maple*. We will list some of them here. Some of the commands are in the **ListTools** package which is not available for *Maple* 6 or earlier versions.

Maple *Command*	*Explanation*

`max (op([ 2, 5, -3, 1.2 , 6.01, 7.5]));`
   7.5
`min (op([ 2, 5, -3, 1.2 , 6.01, 7.5]));`
   -3
	max and **min** are used to find the largest and smallest elements of a list or set, respectively.

`nops ([ 2, 5, -3, 1.2 , 6.01, 7.5]);` 6	**nops** tells you the number of elements in a list or set.

`with(ListTools):`
`Reverse([1, 3, 5, 7, 9]);`
   [9, 7, 5, 3, 1]
	Reverse will reverse the arrangement of elements in a list. That is, the first element becomes the last and vice versa.

`with(ListTools):`
`Rotate([1, 3, 5, 7, 9], 2);`
   [5, 7, 9, 1, 3]
	Rotate will rotate the arrangement of the list.

`with(ListTools):`
`Transpose([[1,3,5,7], [2,4,6,8]]);`
   [[1,2], [3,4], [5,6], [7, 8]]
	Transpose will operates on a list of lists as if it were transposing a matrix. This is particularly useful in changing a pair of lists into a list of ordered pairs.

`with(ListTools):`
`Flatten([[1,3,5,7], [2,4,6,8]]);`
       [1, 3 ,5, 7, 2, 4, 6, 8]
	Flatten combines the elements of both lists into a single list.

More Examples

Baseball Standings

■ **Example.** The following list represents the final standings of the American League Eastern Division Baseball teams for the 1997 regular season in the form [*team*, *wins*, *losses*]:

```
baseball := [ [`Bal`, 98, 64], [`Bos`, 78, 84],
     [`Det`, 79, 83], [`NY`, 96, 66], [`Tor`, 76, 86]]:
```

When you look at the sports page of a newspaper, you will see that the team standings include the teams winning percentage. Also, teams with a higher winning percentage will be listed first. Let us show how to do this in *Maple*:

- We will first add the winning percentage to each element (team):

```
f1 := x -> [op(x[1..3]),     # Note that in this function the element x
     evalf(x[2]/162,3)]:     # is the sublist [team, wins, losses].
```

```
list1 := map(f1, baseball);
```
   list1 := [[*Bal*, 98, 64, .605], [*Bos*, 78, 84, .481], [*Det*, 79, 83, .488],
        [*NY*, 96, 66, .593], [*Tor*, 76, 86, .469]]

- Next we will **sort** the list in descending order of the winning percentage.

```
f2 := (x,y) -> evalb(x[4] > y[4]):
list2 := sort(list1, f2);
```
   list2 := [[*Bal*, 98, 64, .605], [*NY*, 96, 66, .593], [*Det*, 79, 83, .488],
        [*Bos*, 78, 84, .481], [*Tor*, 76, 86, .469]]

- Finally, we will print the contents of each team line by line, just like you see it in a newspaper:

```
for i from 1 to 5 do op(list2[i][1..4]) end do;
```

Bal,	98,	64,	.605
NY,	96,	66,	.593
Det,	79,	83,	.488
Bos,	78,	84,	.481
Tor,	76,	86,	.469

(We will explain the **for** and **do** commands in detail in Chapter 27.)

A U.S. Population Model

■ **Example.** The U.S. population (measured in millions) over the last two hundred years, between 1790 and 1990, is given in ten increments as follows:

```
year :=[seq(1790 + i*10, i=0..20)];
```
   year := [1790, 1800, 1810, 1820, 1830, 1840, 1850, 1860, 1870, 1880,
        1890, 1900, 1910, 1920, 1930, 1940, 1950, 1960, 1970, 1980, 1990]

```
population := [ 3.929, 5.308,  7.240, 9.638, 12.861,
   17.063, 23.192, 31.443, 38.558, 50.189, 62.980,
   76.212, 92.228, 106.021, 123.203, 132.166,
   151.326, 179.323, 203.302, 226.542, 248.710]:
```

Theoretically, this population growth will follow a "logistic model" and hence has the form:

$$\text{population} = \frac{288.5}{1 + e^{a+bx}}, \text{ where } x \text{ is the year}$$

(The constant 288.5 represents the maximum sustainable population 288.5 million, which can be predicted by the data.) We want to find the constants a and b so that the logistic model best fits the data.

We cannot use the **LeastSquares** or **leastsquare** commands (discussed in Chapter 22) to find a and b directly. However, one can easily rewrite the expression $\text{population} = \frac{288.5}{1 + e^{a+bx}}$ as $\frac{288.5}{\text{population}} = 1 + e^{a+bx}$, and thus we have:

$$\ln(\frac{288.5}{\text{population}} - 1) = a + bx.$$

This suggests that we can find a and b using the new data $\ln(\frac{288.5}{\text{population}} - 1)$. Thus, we must transform each of the given data pairs $[x, y]$ to be $[x, \ln(\frac{288.5}{y} - 1)]$.

```
poptrans := map(y->ln(288.5/y-1),population);
```

poptrans := [4.2825978425, 3.9769099964, 3.6596583020,
    3.3650034176, 3.0648925623, 2.7668176917, 2.4370840236
    2.1011214673, 1.8690652898, 1.5577806642, 1.2755916312
    1.0244249128, .75523768025, .54299791023, .29391045440
    .16793603591, –.098186257082, –.49621823061, –.86971459877
    –1.2964736219, –1.8326719347]

```
with(CurveFitting):
LeastSquares(year,poptrans, x);
```

$$55.324474840 - .028552940449\, x$$

This suggests that $a = 55.3245$ and $b = -0.0285529$. The logistic model will thus be:

```
g := x -> 288.5/
        (1 + exp(55.32447484 - .028552940449*x)):
with(stats[statplots]):
pict1 := scatterplot( year, population):
pict2 := plot(g(x),x=1790..1990):
display({pict1, pict2});
```

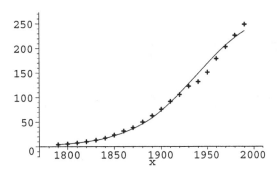

The logistic model fits the data very nicely, especially through about 1930.

Useful Tips

The internal representation of *Maple* lists is different for lists with more than 100 elements. With such long lists we cannot use subscript notation to change values of an element of a list. (For example, you cannot use **longlist[3]** := *new element*.) You may try:

```
newlonglist := [ op(longlist[1..2]), new element,
op(longlist[4..200])];
```

or

```
newlonglist := subsop( 3 = new element, longlist);
```

If order and repetition are not important when working with data, you will probably be more interested in using a *set* to represent the data, rather than in a list. You can then use set commands such as **union**, **intersect**, and **minus**. Recall that sets are written as sequences contained in **{** *curly braces* **}**.

Maple *Command*	*Explanation*
set1 := {2, 5, -3, 1.2, 6.01, 7.5}: **set2 := {1.2, 8, 7.5, 3}:** **set1 union set2;** {-3, 2, 3, 5, 8, 7.5, 6.01, 1.2}	**union** combines the elements of both sets and removes duplicate elements.
set1 intersect set2 ; {7.5, 1.2}	**intersect** finds the elements that are common to both sets and removes duplicate elements.
set1 minus set2 ; {-3, 2, 5, 6.01}	**minus** removes the elements of the first set that are common to both sets.

CHAPTER 25

Random Numbers and Simulation

Random Numbers

The rand Command

We can use *Maple* to generate sequences of random numbers. Such sequences can be used to simulate probabilistic situations.

The following command will generate a random integer between given numbers a and b.

```
rand( a .. b )();
```

If you need a random integer between 0 and $n-1$, you can also enter:

```
rand( n )();
```

For example, you want a random integer between 1 and 6:

```
rand(1..6)();
```
$$6$$

For a random integer between 0 and 9,

```
rand(10)();
```
$$3$$

> **Note:** Notice that there is an extra set of parentheses needed with the **rand** command. This is because the **rand** command will produce a procedure and the extra parentheses are needed to force evaluation of this procedure.

Pseudo-Randomness

Sequences of numbers returned by **rand** are said to be "pseudo-random." They are actually generated using an iteration formula. They eventually repeat but not soon enough so that you would notice.

And, as you see below, the values of **rand** certainly "look" random!

```
datapoints := [seq([i, rand(100)()], i=1..1000)]:
with(plots):
pointplot(datapoints);
```

(A thousand random numbers between 0 and 99 are generated and plotted as the *y*-coordinates.)

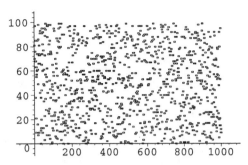

There seems to be no obvious pattern emerging, confirming a sense of randomness.

■ **Example**. Let us generate 5000 random integers between 1 and 10:

```
data := [seq(rand(1..10)(), i=1..5000)] :
```

You can see some of them with:

```
[op(data[1..5]), op(data[4995..5000])];
```
$$[3, 3, 8, 1, 5, 7, 4, 7, 4, 2, 9]$$

If these values were truly random, we would expect each of the numbers 1, 2, . . . , 10 to show up about the same number of times. Let us check it:

```
stats[transform,tally[count]](data);
```

[Weight(1, 497), Weight(2, 490), Weight(3, 513), Weight(4, 500),
 Weight(5, 489), Weight(6, 481), Weight(7, 520), Weight(8, 510),
 Weight(9, 499), Weight(10, 501)]

So, 1 appeared 497 times, 2 appeared 490 times, and so forth. In fact, each digit appeared almost equally often (close to 500 times). Consequently, we say that **rand** produces a uniform distribution of numbers between 1 and 10.

Examples in Simulation

With the help of the **rand** command and a little programming, we can simulate many real-world experiments in *Maple*. In each of the following simulations, usually three procedures are involved.

- We will write down the procedure(s) that generates the result of a single experiment. The procedure is written in the form:

```
game := proc ()
        local local variables separated by commas ;
        Maple commands separated by semicolons or colons ;
        end proc;
```

In many cases the procedure involves the **if** command, used in the form:

```
if condition then result1 else result2 end if;
```

Maple will give *result1* if *condition* is *true*. If *condition* is *false*, *result2* is returned. (The **proc** and **if** commands will be discussed in detail in Chapter 27. In earlier versions of *Maple*, you must use **fi** instead of **end if** to end the **if** statement.)

- Then, we use the **seq** command to repeat the experiment a large number of times in the form:

  ```
  data := [seq( one result of a random experiment, i = 1..n )];
  ```

- After we obtain experimental data in this way, we'll analyze the data and summarize what we observed.

Let's Flip a Coin

■ **Example**. The flipping of a "fair" coin will give us either a head or a tail, with both results being equally probable. That is, approximately 50% of the time we'll get a head, and 50% of the time we'll get a tail. Since **rand(0..1)()** gives a random integer 0 or 1, we can simulate flipping a coin with:

```
flip := proc()  local a;
  a:= rand (0..1)();
  if a = 0  then  "Head"  else  "Tail"  end if;
  end proc:
```

This guarantees that every time *Maple* evaluates **flip()**, you get a new random number and a new flip of the coin.

```
flip();
```

"Head"

```
flip();
```

"Tail"

Here are eight coin flips:

```
seq(flip(), i=1..8);
```

"Tail", "Tail", "Tail", "Head", "Head", "Tail", "Head", "Tail"

Let's Roll a Die

■ **Example**. A standard die has six sides labeled 1, 2, 3, 4, 5, and 6. If the die is "fair," we expect each number to appear on the top face about one-sixth of the time. We can simulate rolling a die with:

```
rand(1..6)();
```

5

```
rand(1..6)();
```

1

Now, suppose that players *A* and *B* each roll a die. The one who gets the larger number wins the game. We can simulate this game:

```
game := proc()
        local player1, player2;
        player1 := rand(1..6)():
        player2 := rand(1..6)():
            if player1 > player2 then `A wins`
            elif player1 = player2 then `Tie`
            else `B wins` end if;

        end proc;
```

Each time you evaluate the command **game()**, you get a new result.

```
game();
```
> *B wins*

```
game();
```
> *Tie*

Let us repeat this game 1000 times and count how many times each player wins.

```
data := [seq(game(), i=1..1000)]:
stats[transform,tally[count]](data);
```

> [Weight(*B wins*, 440), Weight(*Tie*, 167), Weight(*A wins*, 393)]

In this case, it seems that player *B* had better luck.

A Strange Behavior of Numerical Answers

■ **Example**. Look at the answer sections in your physics, chemistry, or economics textbooks. Do you notice that the first digit of each numerical answer is more likely to be 1, 2, 3, or 4 rather than 5, 6, 7, 8, or 9? This is because after each multiplication or division of two numbers, the first digit of the result is more likely to be 1, 2 , 3, or 4 than the others. (Most scientific values arise from multiplications and divisions.)

This phenomenon is related to properties of the logarithm function. If you don't believe it, let us do an experiment to demonstrate it.

We will choose two integers randomly between 1 and 100, multiply them together, and record the first digit of the product. It is tricky to extract the first digit of the product. The idea is to divide the product by a suitable power of 10 such that the answer is a number between 0 and 10. We can then truncate this answer to obtain the first digit of the product. (The truncate command is **trunc**)

```
game := proc()
    local integer1, integer2, theProduct:
    integer1 := rand(1..100)():
    integer2 := rand(1..100)():
    theProduct := integer1*integer2;
    trunc(
    evalf( theProduct /
                10^trunc(log[10](theProduct))))):
    end proc;
```

Here are some test runs:

```
game();
```
> 1

```
game();
```
> 4

Now we repeat this experiment 5000 times.

```
data := [seq(game(), i=1..5000)]:
stats[transform,tally[count]](data);
```

> [Weight(1, 1149), Weight(2, 955), Weight(3, 712), Weight(4, 587), Weight(5, 470), Weight(6, 399), Weight(7, 302), Weight(8, 225), Weight(9, 201)]

In 5000 experiments, we can see that the first digit of the result being 1, 2, 3, or 4 happens 1149+955+712+587 = 3403 times. Are you convinced?

A Game of Risk™, Anyone?

■ **Example**. In the Game of Risk™, players attempt to control a map of the world by occupying countries with their armies. In each turn of the game, a player (the "attacker") may choose to engage another player (the "defender") who occupies an adjacent country in battle. If the attacker eliminates the armies of the defender, the attacker takes over the country.

To simulate a battle, the attacker is allowed to roll two dice, but the defender only one. If the attacker's larger die is higher than the defender's die, the attacker wins the battle. However, the defender wins if his die is higher than or ties the larger die of the attacker. The attacker has the advantage of rolling more dice, but the defender wins ties. So, which player has the advantage?

To find out, we can simulate the action of the attacker by rolling two dice and selecting the largest.

```
attacker := x -> max ( rand(1..6)(), rand (1..6)() ):
```

The action of the defender is easier to simulate:

```
defender := x -> rand(1..6)():
```

A sample battle might look like this:

```
attacker();
                5
defender();
                4
```

In this case, the attacker wins.

The only outcomes of a battle can be that the attacker wins (+1) or loses (−1). Thus we can simulate a battle with:

```
battle := x -> if attacker() > defender() then 1
                    else -1 end if;
```

Now, we can conduct 1000 battles and summarize with:

```
data := [seq( battle(), i=1..1000)]:
stats[transform,tally[count]](data);
```
$$[\text{Weight}(-1, 425), \text{Weight}(1, 575)]$$

The result clearly shows that the attacker has the advantage over time. (In fact, the theoretical probability of the attacker winning over the defender is $\left(\frac{5}{6}\right)^3 \approx 58\%$. Our simulation is quite representative at $575/1000 = 57.5\%$.)

Happy Birthday to YOU!!

■ **Example**. The "birthday problem" is a famous problem in probability. For example, in a room of 30 people, how likely is it that at least two of them have the same birthday? We'll suggest an answer using *Maple*.

It is sensible to assume that birthdays of 30 randomly assembled people are spread evenly over the year, so we can simulate by picking 30 birthdays at random from 365 days of the year (sorry, we don't do leap years in this experiment!)

```
room := proc()
            local i;
            seq(rand (1..365)(), i=1..30);
        end proc;
```

```
room();
```

> 66, 97, 185, 58, 148, 212, 105, 280, 164, 289, 301, 83, 88, 20, 148, 109, 48,
> 166, 327, 214, 208, 168, 275, 3, 277, 233, 288, 308, 65, 1

The output above simulates a typical "room" of 30 people, represented by their birthdays. Notice for this output that the number 148 appears twice, indicating that two people of this group had the same birthday on the 148th day (May 28).

We can count how many *distinct* birthdays there are in the room with:

```
nops({ room() });
```

> 29

(The command **{ room() }** will include the birthdays in a set and hence *removes duplicates*. **nops({ room() })** is then the number of distinct birthdays.) If **nops({room()}) < 30** is *true*, there is a duplicate birthday in the room.

Now we can experiment with 1000 rooms and count the number of *true* results we see:

```
data := [seq(evalb(nops({room()}) < 30), i=1..1000)]:
stats[transform,tally[count]](data);
```

> [Weight(*false*, 298), Weight(*true*, 702)]

That is, the probability that there are two people with the same birthday in a room of 30 randomly assembled people is about 702/1000 = 70.2% Are you surprised? Nearly everyone is surprised the first time they see this result. In fact, the theoretical probability that at least two people have the same birthday in a room of 30 people is approximately 70.6%. Our simulation was very close.

Useful Tips

In this chapter we just gave you a simple idea on how to use random number for simulation. You can find more random tools from the *Maple* package **RandomTools**. (This package is not available for *Maple* 6 or earlier versions.)

When you count the result using **stats[transform,tally[count]]**, your data needs to be expressed in terms of numbers or `names` (using back quotes). You will get an error message if the data contains **"strings"** (using double quotes).

For example, note that we used **stats[transform,tally[count]]** in our "roll a die" and "Game of Risk™" examples, but we cannot use it to summarize results in our "flip a coin" example.

CHAPTER 26

Spreadsheets

Beginning with *Maple* 6, you can include spreadsheets like those found in popular programs such as AppleWorks or Microsoft Excel in your worksheets. We will assume that readers interested in this Chapter are already familiar with the spreadsheet concepts of programs such as these.

Maple spreadsheets can be useful for generating tables for graphic presentation or working with matrices of data. They can also be used to import data from (and export data to) external files.

Basic Spreadsheet Notions

Adding a Spreadsheet

To insert a spreadsheet into a *Maple* worksheet, choose the **Insert → Spreadsheet** menu command. A context bar will appear so that you will be able to work with the spreadsheet:

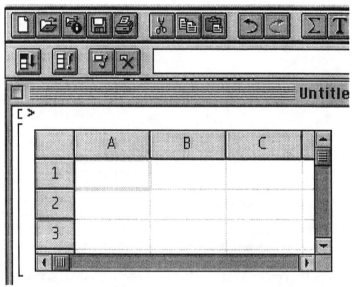

As usual, a spreadsheet is divided into an array of cells, organized into numbered rows and lettered columns. Initially, only the first three rows and columns of a spreadsheet are shown, but you can see more by dragging the resize box in the lower left corner of the spreadsheet. (The spreadsheet is really a resizeable window within a worksheet window.)

You can associate any *Maple* expression with any spreadsheet cell. Simply click on a cell to highlight it and begin typing. Notice that what you type actually appears in the edit field in the Context Bar, and the value of the expression you type will appear in the cell only after you press **Enter** or **Return**.

The following table presents a summary of the basic spreadsheet notions available in *Maple*

Standard Spreadsheet notions	*How they translate to* Maple
Cell Syntax	The tilde character (**~**) is used in formulas to identify cells. For example, cell A1 (column A, row 1) is referred to with the syntax **~A1**. Similarly, cell B5 would be referred to as **~B5**.
Auto Fill feature	The Fill button ▣ appears on the left of the context bar. If a cell is selected when you click this button, you can choose both the direction of the fill (up, down, left or right) and an optional step size for the fill. Formulas in cells are filled using *relative* cell references, as explained in the next item. (You can also use the **Spreadsheet** → **Fill** menu command.)
Relative Cell References	The syntax **~A1** denotes a *relative* reference. If you use, say, the formula (**~A1**)**^2** in cell B1, it means "the square of what is in the cell to the left of this cell." If you fill down from cell B1 to B2, then the formula for B2 will be entered as (**~A2**)**^2**, which again means "the square of what is in the cell to the left of this cell."
Absolute Cell References	Use a dollar sign (**$**) in a formula (as you would in any other spreadsheet program) when you want to reference a fixed cell by row or column. For example, the syntax **~A1** will always reference cell A1, no matter into which cell it is copied or filled.

A Table of Derivatives and Integrals

■ **Example.** We will use a spreadsheet to construct a table of derivative and integral formulas.

Use the **Insert** → **Spreadsheet** menu command to create a spreadsheet. Drag the resize box in its lower left corner to reveal 6 rows and 3 columns.

- Across the first row, type **f(x)**, **Diff(~A1,x)**, and **Int(~A1,x)** as the entries of the first row. (Notice that **Diff** and **Int** are the inert differentiation and integration commands.)

- Now type the expressions **x^n**, **sin(x)**, **tan(x)**, **ln(x)**, and **arcsin(x)** in cells A2 through A6.

- Type **simplify(diff(~A2,x))**, and **simplify(int(~A2,x))** in cells B2 and C2.

The result should look like this:

	A	B	C
1	$f(x)$	$\dfrac{\partial}{\partial x}f(x)$	$\int f(x)\,dx$
2	x^n	$x^{(n-1)}\,n$	$\dfrac{x^{(n+1)}}{n+1}$
3	$\sin(x)$		
4	$\tan(x)$		
5	$\ln(x)$		
6	$\arcsin(x)$		

Now click and drag to select cells B2 through C6. From the **Spreadsheet** menu select **Fill** → **Down** to obtain the desired table of formulas. Congratulations … you now have a very cool-looking table of derivatives and integrals!

	A	B	C
1	$f(x)$	$\dfrac{\partial}{\partial x}f(x)$	$\int f(x)\,dx$
2	x^n	$\dfrac{x^n\,n}{x}$	$\dfrac{x^{(n+1)}}{n+1}$
3	$\sin(x)$	$\cos(x)$	$-\cos(x)$
4	$\tan(x)$	$1+\tan(x)^2$	$-\ln(\cos(x))$
5	$\ln(x)$	$\dfrac{1}{x}$	$x\ln(x)-x$
6	$\arcsin(x)$	$\dfrac{1}{\sqrt{1-x^2}}$	$x\arcsin(x)+\sqrt{1-x^2}$

Properties of Spreadsheets

Every spreadsheet you insert into a worksheet has properties associated with it, including its name, cell alignment, numeric display precision, and background color.

If you click on a spreadsheet, you can use the **Spreadsheet** → **Properties** menu command to find out information about a spreadsheet. As a shortcut, you can right-click (or Mac option click) on a spreadsheet and a Context menu pops up from which you can select the **Properties...** menu item.

The most important property of a spreadsheet is its name. By default, the first spreadsheet you create in a *Maple* session is named "SpreadSheet001," the second "Spreadsheet002," and so on. You can change the name of a spreadsheet using the **Spreadsheet** → **Properties** menu command to be more meaningful.

The Spread Package

The **Spread** package provides a number of commands so that you can programmatically access and work with spreadsheets.

```
with(Spread);
```

[CopySelection, CreateSpreadsheet, EvaluateCurrentSelection, EvaluateSpreadsheet, GetCellFormula, GetCellValue, GetFormulaeMatrix, GetMaxCols, GetMaxRows, GetSelection, GetValuesMatrix, InsertMatrixIntoSelection, IsStale, SetCellFormula, SetMatrix, SetSelection]

These commands let you write *Maple* programs that can: create a spreadsheet; access data and/or change the formula associated with any cell; manipulate the current cell selection; and move data back and forth between a spreadsheet and a matrix.

Input a Matrix Using a Spreadsheet

■ **Example**. If you need to enter a (large) matrix of data in a *Maple* session, it can be tedious to specify the rows and columns properly using the syntax of matrices. But you can enter the data more easily in a spreadsheet, and then move the data into a matrix.

Computer scientists and those studying Discrete Mathematics will recognize that the graph shown below (on the left) can be represented by the matrix shown below (on the right), called its adjacency matrix:

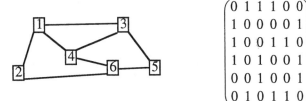

$$\begin{pmatrix} 0 & 1 & 1 & 1 & 0 & 0 \\ 1 & 0 & 0 & 0 & 0 & 1 \\ 1 & 0 & 0 & 1 & 1 & 0 \\ 1 & 0 & 1 & 0 & 0 & 1 \\ 0 & 0 & 1 & 0 & 0 & 1 \\ 0 & 1 & 0 & 1 & 1 & 0 \end{pmatrix}$$

We can avoid many common input errors in describing this matrix in *Maple* by using the spreadsheet interface. Begin by creating a new spreadsheet in your worksheet with the command:

```
with(Spread):
CreateSpreadsheet("adjMatrix");
```

Now enter the data of ones and zeroes for the matrix shown above into the spreadsheet, in the upper, left corner of the spreadsheet (cells A1 through F6). Once the data are entered in the spreadsheet, select the cells in the spreadsheet.

	A	B	C	D	E	F
1	0	1	1	1	0	0
2	1	0	0	0	0	1
3	1	0	0	1	1	0
4	1	0	1	0	0	1
5	0	0	1	0	0	1
6	0	1	0	1	1	0

Now you can move the data into a matrix **m** with the command:

m := GetValuesMatrix("adjMatrix");

$$m := \begin{pmatrix} 0 & 1 & 1 & 1 & 0 & 0 \\ 1 & 0 & 0 & 0 & 0 & 1 \\ 1 & 0 & 0 & 1 & 1 & 0 \\ 1 & 0 & 1 & 0 & 0 & 1 \\ 0 & 0 & 1 & 0 & 0 & 1 \\ 0 & 1 & 0 & 1 & 1 & 0 \end{pmatrix}$$

It is now easy to work with the matrix **m**. For example, higher powers of **m** provide information about the length and number of paths between nodes in the graph. We can see that it is possible to travel from any one node in the graph to any other node in the graph in no more than 3 steps, since the third power of the matrix contains no zeroes off the diagonal:

m.m.m;

$$\begin{pmatrix} 2 & 5 & 5 & 6 & 3 & 2 \\ 5 & 0 & 3 & 1 & 1 & 5 \\ 5 & 3 & 2 & 6 & 5 & 2 \\ 6 & 1 & 6 & 2 & 1 & 7 \\ 3 & 1 & 5 & 1 & 0 & 5 \\ 2 & 5 & 2 & 7 & 5 & 0 \end{pmatrix}$$

Importing/Exporting Data with Spreadsheets

A useful feature of a spreadsheet is that its contents can be populated from an external datafile, and its values can be exported to an external file.

The external formats that are understood by *Maple* include tab-delimited data (perhaps used by a stand-along spreadsheet program), as well as MatrixMarket and Matlab files. When you combine this capability with the matrix conversion commands in the **Spread** package, you have a very nice way to preview and edit data when reading it from (writing it to) an external file.

Exporting Data using a Spreadsheet

Exporting data from a spreadsheet is done with the **Spreadsheet → Export Data** menu command.   All *values* in the spreadsheet are exported to the type of file you choose (the *formulas* of the cells are not exported, although they will be saved in your *Maple* worksheet).

For example, we can use this capability to save values from a *Maple* matrix to an external file, while being able to view and edit any of the entries before actually saving the data.

The following sequence places the contents of a matrix named **m1** into a new spreadsheet and you can examine its values:

```
with(Spread);
CreateSpreadsheet(mySheet1);
SetMatrix(mySheet1, m1, 1, 1);
EvaluateSpreadsheet(mySheet1);
```

You can now see the values and use the spreadsheet interface to edit any of them if you like. Once you're happy with the values and are wish to write them to an external file, select the cells of the spreadsheet you wish to write and use the **Spreadsheet → Export Data** menu command. (*Only the selected portion of the spreadsheet is written to the external file, so it is easy to choose any sub-matrix of values to write to a file.*)

Importing Data using a Spreadsheet

Conversely, if data are stored in a file, you can bring the values into *Maple* using a spreadsheet, examine/edit those values, and then move any selection of values in the spreadsheet into a matrix for further *Maple* processing.

After inserting a spreadsheet into your *Maple* worksheet, use the **Spreadsheet → Import Data** menu command, and select both the file type and the file from which you will import the data.

Next, you must use the **Spreadsheet → Evaluate Spreadsheet** menu command (or click the button), because the data read in are treated as strings or formulas, and this command forces the spreadsheet to evaluate the strings.

After examining the data in the spreadsheet -- and possibly editing any entries if necessary -- select the portion of the spreadsheet you wish to move into a *Maple* matrix (named **m2** below) and execute the sequence:

```
with(Spread);
m2 := GetMatrixValues("SpreadSheet001");
```

Useful Tips

The spreadsheet features supplied within *Maple* are very limited. Most of the built-in spreadsheet functions you'd expect to find in a true spreadsheet program (e.g., computing the sum or average of a row or column) are *not* present in *Maple*. In particular, the notion of specifying a range of cells in a formula is missing from *Maple*.

We think that the primary use of *Maple* spreadsheets will be via programmatic access, usually for the purpose of editing matrix entries and viewing/editing data as you move it to and from external files.

Troubleshooting Q & A

Question... When I tried using **SetMatrix** or **SetCellFormula** command to assign a spreadsheet formula that depends on another cell, an unexpected value is put in place. What went wrong?

Answer... Unless you say otherwise, *Maple* will evaluate the expression you are using as a formula before assigning it to the cell. Enclosing an expression in single quotes delays evaluation of it.

Question... When copying and pasting in a spreadsheet, I got unexpected results. What went wrong?

Answer... *Maple* uses relative references when copying and pasting formulas. The original cell addresses that your formula used have been changed.

Question... I am having trouble with the spreadsheet function **GetValuesMatrix**. I keep getting the wrong values selected in the spreadsheet. What happened?

Answer... There was a bug in *Maple* 6 (corrected in *Maple* 7 and later versions, however) in which the **GetValuesMatrix** command would return the correct number of rows and columns for a selection, but only starting at column A and row 1.

You can work around this bug -- enlarge the selection to include cell A1, then use the **LinearAlgebra** package routine **SubMatrix** to extract the submatrix at the proper position in your spreadsheet.

CHAPTER 27
Maple for Programmers

This chapter is designed for *Maple* users who are already familiar with writing computer programs. However, it is not a tutorial on programming. Rather, we expect that you are looking for information on those areas of *Maple* that are closely related to programming concepts.

Elements of Traditional Programming Languages

Maple has its own programming language that supports elements found in traditional computer programs: conditional execution, looping, and subroutines. The syntax of the language borrows from the Pascal and C programming languages.

> **Note:** Until now, we have used the term "command" to describe *Maple* syntax (e.g., we used the **plot3d** command in Chapter 15). In this chapter, most commands will be called **statements**. This is more consistent with the terminology of programming languages.

The print Statement

To print the value of one or more expressions, use the statement:

```
print( expression1, expression2, ... );
```

When you are trying to debug programs it is useful to add **print** statements to let you see what the program is actually doing. You can also use the **printf** command to format the output. (You can type **?printf** to see the details.)

The if Statement

The most commonly used form of the **if** statement is:

```
if logical condition then result1; else result2; end if;
```

(Notice that the **if** statement ends with **end if**. In earlier versions of *Maple* the **if** statement had to end with **fi**, a backwards **if**.) If the *logical condition* evaluates as *true*, *result1* is returned. If the *logical condition* evaluates as *false*, *result2* is returned. For example:

```
a := 4:  if  a <= 3 then 2*a-1; else 5*a+1; end if;
              21

a := 3:  if  a <= 3 then 2*a-1; else 5*a+1; end if;
              5
```

In an **if** statement, each of *result1* and *result2* can be either a single statement or a collection of statements separated by semicolons. The logical decision can also have more than two branches by using the optional **elif** clause (for "else if"):

```
if logical condition then result1;
elif logical condition then result2;
                    else result3;
end if;
```

Consider, for example, a procedure to solve the quadratic equation $ax^2 + bx + c = 0$:

```
a := 4:      b := 7:    c := 2:
disc := b^2 -4*a*c:
if disc > 0 then
   print(`The equation has two solutions at`);
   print((-b+sqrt(disc))/(2*a),`  and  `,
         (-b-sqrt(disc))/(2*a));
elif disc = 0 then
   print(`The equation has a double solution at`);
   print((-b)/(2*a));
else
   print(`The equation has no real solutions`);
end if:
```

The equation has two solutions at

$$-\frac{7}{8} + \frac{1}{8}\sqrt{17}, \text{ and }, -\frac{7}{8} - \frac{1}{8}\sqrt{17}$$

Do Loop

The simplest kind of loop is the **do** statement. The syntax is:

```
do body end do;
```

(In earlier versions of *Maple*, the **do** statement had to end with **od** instead of **end do**. **od** is **do** backwards.) The *body* can be either a single *Maple* statement or a sequence of statements separated by semicolons. To keep the loop from executing forever, you can use the **break** statement within the *body* of the **do** statement.

■ **Example.** The **nextprime** command can be used in a **do** loop to make a short list of primes:

```
aprime := 2;
do
    aprime := nextprime(aprime);
    if aprime > 12 then break end if;
end do;
```

```
aprime := 2
aprime := 3
aprime := 5
aprime := 7
aprime := 11
aprime := 13
```

Using a **do** loop with one or more **break** commands to stop repetition is rather inelegant. The **for** loop provides more control over repetition of the loop.

For Loop

The most common **for** loop has the form:

```
for index variable from starting value to stopping value
    do  body  end do;
```

■ **Example.** The Fibonacci numbers are $c_0 = 1$, $c_1 = 1$, and $c_n = c_{n-1} + c_{n-2}$, for $n \geq 2$. Executing the assignment **c[n] := c[n-1] + c[n-2]** repeatedly for $n \geq 2$ computes these directly:

```
c:= array(0..25):
c[0] := 1:
c[1] := 1:
for i from 2 to 25 do
   c[i] := c[i-1] + c[i-2];
end do:
```

To see a few of the Fibonacci numbers, you can use:

```
print(seq(c[i],i=0..10));
```

$$1, 1, 2, 3, 5, 8, 13, 21, 34, 55, 89$$

```
c[21];
```

$$17711$$

Extensions of the For Statement

The general form of the **for** statement is this (parenthetical terms are optional):

> **for** *index variable* (**from** *expression*)
>   (**by** *expression*) (**to** *expression*)
>   (**while** *expression*) **do** *body* **end do;**

If **from** and **by** are omitted, the default values of **from 1** and **by 1** are used. The **to** expression and **while** expression are tested at the beginning of each loop. Both can be used at the same time. The index variable is increased using the **by** value at the end of the loop.

■ **Example.** To compute the sum of the positive integers $1, 2, 3, \ldots, 50$ using the **for** statement, use:

```
intsum := 0:
for i from 1 to 50 do intsum := intsum + i end do:
intsum;
```

$$1275$$

To compute the same sum using the **while** statement, use:

```
intsum := 0:
for i from 1 while i <= 50 do intsum := intsum + i
end do: intsum;
```

$$1275$$

(Of course, one can also get this answer easily using **add(i, i=1..50);**)

Procedures, Functions, and Subroutines

Almost every programming language supports a notion of a subroutine, function, or procedure. For example, BASIC allows the use of **DEF**, **GOSUB**, and **RETURN** statements, while FORTRAN has a **CALL** statement.

Subroutines are best written in *Maple* using the **proc** statement. Its general form is:

```
proc( parameter list)
   local local variable list;
   global global variable list;
   options options list;
        body
end proc;
```

Maple will execute the statement(s) in the *body* and then return the value of the last statement executed. Here are two examples.

■ **Example.** To compute the area of a triangle with sides having length a, b, and c, you can use Heron's Formula which says that $area = \sqrt{s(s-a)(s-b)(s-c)}$, where $s = (a+b+c)/2$ is the semi-perimeter of the triangle. A nice coding of an area function (subroutine) would be:

```
area := proc(a,b,c)
        local s;
        s := (a+b+c)/2;
        sqrt(s*(s-a)*(s-b)*(s-c));
        end proc:

area(5, 5 ,8);
        12
```

> **Note:** Don't forget to add semicolons or colons between statements in the body of a **proc**. Otherwise, *Maple* stops with a syntax error.

Since the length of each side should be positive, we can modify the procedure to work only if all three of the input parameters are positive.

```
area := proc(a:: positive,b::positive,c::positive)
        local s;
        s := (a+b+c)/2;
        sqrt(s*(s-a)*(s-b)*(s-c));
        end proc:

area(-2,1,5);
```

Error, area expects its 1st argument, a, to be of type positive, but received -2

■ **Example.** We can write a routine that computes the maximum and average of a list of exam scores from a Calculus class, and shows the scores in increasing order.

```
summarize := proc(data)
   local sortedList, n:
   sortedList := sort(data):
   n := nops(data);
   print(`The number of students is `, n);
   print(`The maximum score is `, sortedList[n]);
   print(`The average is `,
           evalf(sum(data[k], k=1..n)/ n, 4));
   sortedList;
end proc:

class1 := [61,23,14,78,91,33,12,44,72,79,82,81];
summarize(class1);
```

The number of students is 12
The maximum score is 91

The average is 55.83
[12, 14, 23, 33, 44, 61, 72, 78, 79, 81, 82, 91]

> **Note:** Printing results from an operation on the inside of a nested loop is suppressed unless a **print** statement is used or the **printlevel** variable is changed to a higher number.

Modules

Starting with *Maple* 6, the programming structure of a module is available for large programs and packages. Module programming is more object-oriented. However, this topic is beyond the scope of this book.

Recursive Structures in *Maple*

Maple allows recursion, which means that a procedure can call itself.

■ **Example.** The Fibonacci numbers are defined recursively as $c_0 = 1$, $c_1 = 1$, and $c_n = c_{n-1} + c_{n-2}$, for $n \geq 2$. These can be defined in *Maple* with:

```
c := proc(n::nonnegint)
    if n < 2 then 1
    else c(n-1) + c(n-2)
    end if;
end proc:
```

The procedure call checks that **n** is a nonnegative integer. The condition **if n < 2 then 1** will define **c(0) := 1** and **c(1) := 1**. To find the Fibonacci number c_5, just type:

```
c(5);
```

                8

To evaluate **c(5)**, *Maple* repeatedly uses the three rules above to simplify the expression in approximately the following sequence of transitions:

$$c_5 \to c_4 + c_3 \to (c_3 + c_2) + (c_2 + c_1) \to ((c_2 + c_1) + (c_1 + c_0)) + ((c_1 + c_0) + 1)$$

$$\to (((c_1 + c_0) + 1) + (1+1)) + ((1+1) + 1) \to (((1+1) + 1) + 2) + (3) \to 8$$

Remembering Values During Recursion

One disadvantage of the method above is that *Maple* will not remember the values that it has calculated recursively. For example, if you want to calculate **c(6)**, *Maple* recomputes **c(5)**, **c(4)**, **c(3)**, and **c(2)** all over again. This can slow down computation dramatically even for just, say, **c(20).**

The **remember** option of *Maple* gives you a way to evaluate **c(5)** and at the same time define its value in case you need to use it later. You do this with the following new definition of the Fibonacci procedure:

```
newFib := proc(n::nonnegint)
    option remember:
    if n < 2 then 1
    else newFib(n-1) + newFib(n-2)
```

```
    end if;
end proc:
```

This option makes *Maple* create a "remember table" so that **newFib(i)** is only computed once for each input value of **i**. On subsequent calls, **newFib(i)** is evaluated with a lookup table.

Using **remember** has a dramatic effect on execution time. Evaluating **c(20)** using the first recursive method above takes *two to three hundred* times longer than **newFib(20)** of the second method! You can compare the CPU time used in these two computations with the **time** command.

time(c(20));   #The output depends on how fast is your machine.

.149

time(newFib(20));

0.

Code Generation

Fortran and C

After we have worked with a procedure in *Maple*, we may want to use the results in a different computer program. This can be done with commands from the **codegen** package. The *Maple* commands **fortran** and **C**, respectively, produce FORTRAN or C code equivalents of expressions and procedures.

■ **Example.** The procedure **newFib** above for finding Fibonacci numbers includes an error checking option that is not available in either FORTRAN or C. We modify the *Maple* procedure slightly and then let *Maple* produce the FORTRAN or C code for us.

```
newFib := proc(n)
   option remember:
   if type(n,negint) then
        error("bad arguments type")
   elif n < 2 then 1
            else newFib(n-1) + newFib(n-2)
   end if;
end proc:

with(codegen):
C(newFib);
/* The options were    : remember */
double newFib(n)
double n;
{    if (type(n,negint)) {
          fprintf(stderr, "bad arguments type" );
          exit(1);
          }
     else if (n < 2)
          return(1);
     else
          return(newFib(n-1)+newFib(n-2));
}

fortran(newFib);
```

(We will not show you the output here.)

Both of these commands have options to optimize the code and save the results directly to a file.

Java and Code Generation

Maple 8 introduced a new package, **CodeGeneration** that will eventually replace the **codegen** package.

Of interest to us is that it generates not only FORTRAN and C, but Java code as well. The new package more explicitly points out which *Maple* structures do not translate into the target language.

■ **Example.**   The procedure **newFib** above for finding Fibonacci numbers uses a remember table for increased efficiency. This structure does not translate into FORTRAN, C, or Java. The **codegen** package produced code with a comment that indicates the "remember option" is not translated. When we try generating Java code with **CodeGeneration** the result is clearer.

```
with(CodeGeneration):
Java(newFib);
```

Warning, procedure options ignored
Error, (in IssueError) cannot analyze remember tables

If we delete the remember option, *Maple* will generate code, but it will inform us that the function **type** is not defined in the target language. Instead of using the **type** command, we can check directly to see if the input is less than zero.

```
newFib2 := proc(n)
   if n < 0 then
       error("bad arguments type")
   elif n < 2 then 1
   else newFib2(n-1) + newFib2(n-2) end if; end proc;
Java(newFib2);
```

```
class CodeGenerationClass {
 public static int newFib2 (int n)
 {
  int cgret;
  if (n < 0) {
   System.err.println("bad arguments type");
   System.exit(1);
  }
  else if (n < 2)
   cgret = 1;
  else
   cgret = newFib2(n - 1) + newFib2(n - 2);
  return(cgret);
 }
}
```

The **FORTRAN** and **C** commands in the **CodeGeneration** package behave similarly. *Maple* also allows you to make calls to external code written in FORTRAN, C, and Java, but that feature is beyond the scope of this book.

Maplets Package

With *Maple* 8 and higher, users can create Maplets, which bring a Java interface to a *Maple* program. Examples of Maplets are available at the *Maple* Application center www.mapleapps.com. Maplets are created with the tools in the **Maplets** package. Maplet programming is beyond the scope of this book.

Viewing the Code of *Maple* Procedures

Almost all of the statements (commands) in *Maple* are written in the *Maple* programming language. This code can be viewed on demand. You may want to do this to see exactly how *Maple* executes a command, or because you want to modify or extend a procedure.

The showstat Command

The easiest way to view *Maple* code is with the **showstat** (show statements) command. For example, you can see the code for the *Maple* command **norm**:

```
showstat(norm);
```

```
norm := proc(p, n, v)
local c;
   1   c := traperror(coeffs(p,args[3 .. nargs]));
   2   if c = lasterror then
   3     error "polynomial must be expanded"
       elif n = infinity then
   4     max(op(map(abs,[c])))
       elif n = 1 then
   5     convert(map(abs,[c]),`+`)
       elif n = 2 then
   6     `norm/ex`(convert(map(proc (x) options operator, arrow; abs(x)^2 end proc,[c]),`+`),1/2)
       elif type(n,'numeric') and 1 <= n then
   7     `norm/ex`(convert(map(proc (x, n) options operator, arrow; abs(x)^n end proc,[c],n),`+`),1/n)
       else
   8     error "norm not implemented"
       end if
end proc
```

Another way of showing code in a *Maple* procedure is to change the value of the system variable **verboseproc** with the **interface** command. This variable is usually set at **1**. To see code in a procedure the variable should be set to **2**. Setting the variable to **3** will let you see remember tables as well. Use the print command to see the code. For example to see the code for **newFib** that we defined above.

```
interface(verboseproc=3);
print(newFib);
```

(We will not show the output here.)

Debugging Code

To understand what a block of code does, it is useful to trace through values of the assigned variables as the procedure executes. The option to trace a procedure is turned on with the **debug** command and turned off with the **undebug** command.

```
debug(newFib):
newFib(10);
undebug(newFib);
```

(We will not show the output here.)

You should be warned that most common commands are composed of several pages of code, as *Maple* carefully breaks most commands into many cases.

Another useful command for debugging procedures is **printlevel**. In *Maple*, statements within a procedure are recognized in levels, determined by nested loops, conditional or repetition statements. By resetting the **printlevel** to different levels, *Maple* will then display the results of all statements executed up to that level.

You can see the difference in the output by changing the **printlevel** in the following statements.

```
printlevel := 2;   # also try levels 1 and 3 to compare
  a := 1; b :=1; c:=1;
  if a <3 then
     if b < 3 then
        if c < 3 then
           d:= a+b+c;
           c := c+1 end if;
        b := b+1 end if;
     a := a+1 end if;
```

(We will not show the output here.)

File I/O

Saving Values or Procedures for Future *Maple* Use

You can **save** results from one session of *Maple* and then **read** them back in during a future *Maple* session. Both commands use a pathname that starts in the same directory as the *Maple* application. The statement

```
save(newFib, a, "mondayMaple.m");
```

saves the variable **a** and the procedure **newFib** in a file named **mondayMaple.m**. The file would be read back in during a later *Maple* session with

```
read("mondayMaple.m");
```

This is an internally formatted *Maple* file and is not readable with a text editor. It is, however, in ASCII so that it can be transferred via e-mail.

Importing Data

To import a data file to *Maple*, you will use the **readdata** command. Suppose you have a text file called **file1.txt** with the data one to a line. You can use the following command to download the data:

```
readdata("file1.txt");
```

Exporting Data

If you have a small amount of numeric data to export to a file, you can use the spreadsheet feature (*Maple* 6 and higher) to do this rather quickly and easily (see Chapter 26).

However, to export either a large amount of numeric data from *Maple*, or to do so programmatically, you must go through a three-step process of opening a file, writing to it, and closing the file.

• First, you must create file descriptor for the output file. For a buffered file, you use the **fopen** command in the form:

```
fopen( file name, READ/WRITE/APPEND, TEXT/BINARY );
```

For example, if we want to append text to **file2**, we would start with the statement:

```
fd1 := fopen("file2", APPEND, TEXT);
```

- Next, you use the **writedata** statement to output the values. You may write a list or vector or matrix. To write **myData** to **file2**, use the statement:

```
myData := [20, 31, 15, 26, 17, 18];
writedata(fd1, myData);
```

- Finally, when you are finished with output, you should **fclose** the file:

```
fclose(fd1);
```

If you use a text editor to open **file2**, you will see the data 20, 31, etc., one to a line. You can read this file back to *Maple* using the **readdata** command.

```
readdata("file2");
```

$$[20., 31., 15., 26., 17., 18.]$$

If you want to write the data 20., 31., etc. all on one line, you must enter them as a list of a list:

```
fd2 := fopen("file3", APPEND, TEXT):
myData2 := [[20, 31, 15, 26, 17, 18]];
writedata(fd2, myData2);
fclose(fd2);

readdata("file3");    #read the 1st element of each line
```

$$[20.]$$

```
readdata("file3", 2);  #read the first 2 elements of each line
```

$$[[20., 31.]]$$

Optionally, you can add formatting information after the data. More than one **writedata** statement can be used between the **fopen** and **fclose**.

■ **Example.** Suppose that you want to compute the values of the Fibonacci numbers with the procedure **newFib** defined above, and then save those values to a file named **"Fibonacci.dat"**, first with all 25 values on one line, then with 25 lines each containing the list **[i, newFib(i)]**.

```
fd1 := fopen("Fibonacci.dat",WRITE,TEXT):
writedata(fd1,[[seq(newFib(i),i=1..25)]]);
for i from 1 to 25 do
    writedata(fd1, [[i,newFib(i)]]);
end do;
fclose(fd1);
```

To read the first column of each line we can use the **readdata** command:

```
readdata("Fibonacci.dat");
```

APPENDIX A
Worksheets

Worksheets and Groups

Worksheets

Worksheets are the basic interface elements of *Maple*. You enter all your *Maple* input in a worksheet, and view all of *Maple*'s output including graphics in a worksheet.

Execution Groups

Each worksheet is composed of **execution groups**. The groups are indicated by "group brackets" that appear on the left side of the worksheet window, as shown here:

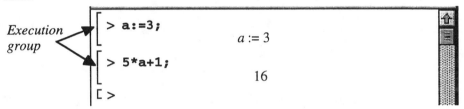

The two execution groups indicated above contain the input and output regions. *Maple* will automatically combine each input and output to form an **execution group**. In general, execution groups are formed from input, output, text and graphics regions.

Inserting a New Execution Group

After each computation, *Maple* will automatically create a new execution group waiting for your input. However, you can also insert a new group at any space between two existing groups. Let us show you how to do so.

Say you want to add a computation before the input **5*a+1**. There are two ways to do so:

- Method 1: Place the cursor in the output $a := 3$. Press the ⬛ button in the tool bar. A new execution group will be created before the input **5*a+1**.

- Method 2: Place the cursor in the input **5*a+1**. On the menu bar, choose the **Insert** menu and select **Execution Group** → **Before Cursor**. A new execution group will be created before the input **5*a+1**.

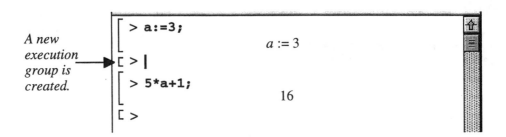

A new execution group is created.

```
[ > a:=3;
                           a := 3
[ > |

[ > 5*a+1;
                           16
[ >
```

Inserting Text Between Execution Groups

Instead of entering an input command, you can also create a **text region**. This allows you to add commentary, titles, or headings to your work.

Suppose you want to add some commentary before the input **5*a+1**. To do so,

- Following the previous instruction (method 1 or 2), insert a new execution group before the input **5*a+1**.

- Press the **T** button in the tool bar. ("T" stands for text.)

- The context bar will change to a text menu. Choose the text style and font. You can start typing your commentary

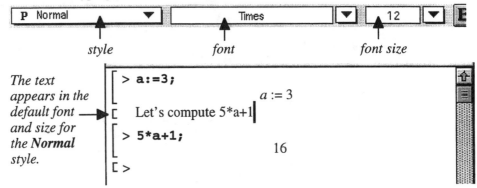

style	*font*	*font size*

*The text appears in the default font and size for the **Normal** style.*

```
[ > a:=3;
                           a := 3
[
[    Let's compute 5*a+1|

[ > 5*a+1;
                           16
[ >
```

Using a similar procedure, you can also change the text style to **Title, Author, Bullet Item**, and so on. Each of these styles has its own format. You can experiment with them.

Editing Groups and Text

The following table summarizes the editing procedures you will use the most when working in a worksheet:

What You Want to Do	*How to Do It*
Delete an execution group.	• Highlight the group bracket at the left. • Choose **Delete Paragraph** from the **Edit** menu.
Make a copy of an execution group in a new location.	• Highlight the group bracket at the left. • Choose **Copy** from the **Edit** menu. • Create a new execution group at the location and use either the **[>** button in the tool bar or the **Execution Group** in the **Insert** menu. • Choose **Paste** from the **Edit** menu.

Cut, copy, or paste the text of a group within the same or another group.	• Handle this the same way that you manipulate text in any word processor. (Use the mouse to select, and then use **Cut**, **Copy**, or **Paste**.)
Change the font, size, or style of some (or all) of the text in a text region.	• Select the text with the mouse. • Choose the appropriate font, size, and style from the text menu context bar.
Change the default font, size, or style.	• In the menu bar, choose the **Format** → **Style** menu. Select the appropriate font, size, and style.

Worksheet Organization

Titles, Sections, and Subsections

A well-organized document has a title, perhaps a subtitle, and is divided into a number of sections. The sections may further be broken into subsections.

You can structure a *Maple* worksheet in much the same way by using the **Title** style and the **Indent** feature. By dividing the worksheet into sections, you will produce a nicely organized document that is suitable for both reading and printing.

■ **Example.** Suppose we have made the following simple calculus computations in a worksheet:

```
> limit(sin(x)/x, x=0);
                    1
> diff(sin(x), x);
                  cos(x)
> diff(x^2+2*x, x);
                  2 x+2
```

To organize these results, we will add a title named "Math examples." This can be done by choosing the **Execution Group** → **Before Cursor** in the **Insert** menu, clicking the 🅣 button in the tool bar, and choosing the **Title** style as we mentioned in the last section.

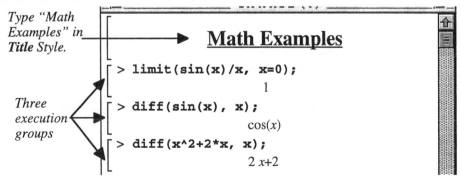

Type "Math Examples" in Title Style.

Three execution groups

Math Examples

```
> limit(sin(x)/x, x=0);
                    1
> diff(sin(x), x);
                  cos(x)
> diff(x^2+2*x, x);
                  2 x+2
```

Now we want to add a section heading to say that these are from "Calculus."

- First, select the three execution groups (see the picture above) by highlighting all their input and output regions.
- In the menu bar, select the **Format** menu and choose **Indent**. A large bracket topped by a little square will appear to the left.
- Move the cursor next to the ▣, and type "Calculus." (See the picture below.)

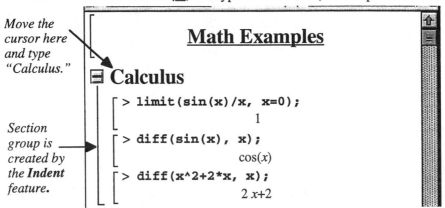

Move the cursor here and type "Calculus."

*Section group is created by the **Indent** feature.*

Similarly, you can add two subsection headings that separate the material into "Limits" and "Derivatives."

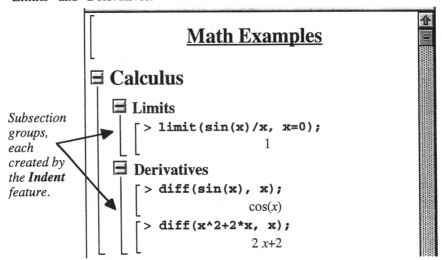

*Subsection groups, each created by the **Indent** feature.*

Open and Closed Sections

All the sections of the previous window are said to be open, because you can see their contents. You can close a section or subsection so that you only see its heading. This provides a very nice outlining capability, as you will see.

For example, the "Derivatives" subsection can be closed by clicking its ▤ button.

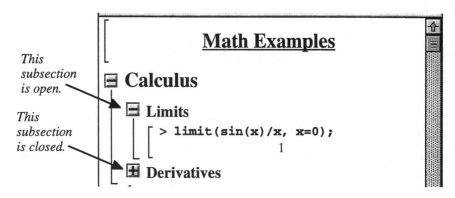

This subsection is open.

This subsection is closed.

If you also close the "Limits" subsection, the window becomes even more compact.

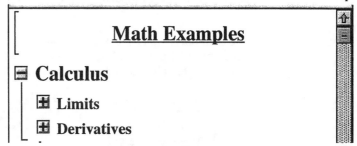

To open the section or subsection, click on the corresponding ⊞ button.

**Printing a
Worksheet**

You can print your worksheet from **Print** in the **File** menu. If you want to hide the group and section brackets, you can de-select the **Show Section Ranges** and **Show Group Ranges** from the **View** menu. This will produce a much more attractive printed document.

Input Shortcuts

Input

In *Maple* there are two shortcuts to enter commands and expressions. They are especially useful for preparing worksheets for printing and publication.

Palettes

In the menu bar, under the **View** menu, you can find the **Palettes** command. There are four types of palettes: **Expression Palette**, **Matrix Palette**, **Symbol Palette** and **Vector Palette**. (*Maple* 6 does not have the **Vector Palette**.)

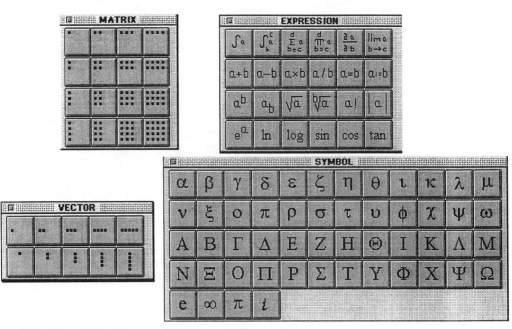

We will quickly illustrate how they work.

■ **Example**. To evaluate $\int x^2 \, dx$ you can type:

```
int(x^2, x);
```

or, you can use the **Expression** palette:

• Open the **Expression** palette by selecting **Palette** → **Expression Palette** under the **View** menu.

• Choose **Standard Math Input** under the **Insert** menu on the menu bar. You will see a highlighted **?** next to the *Maple* input prompt **>**.

- Click on the $\int_{\square}$ button in the **Expression** palette. A new expression is created in your worksheet that looks like:

$$\int ? \, d?$$

- Click on the $a^{\flat}$ button. Now you see:

$$\int ?^{?} \, d?$$

- Type *x*, then *tab*, then 2, then *tab*, and then *x* again. Every time you press the *tab* key, you move from one box in the expression to another box. Now you see the completed expression:

$$\int x^{2} \, dx$$

- Hit the enter key to evaluate. You now get the expected result:

$$\frac{1}{3} x^{3}$$

The Context Menu

There is a shortcut to differentiate, integrate, factor, or evaluate a previous *Maple* output. Consider the result $\frac{1}{3} x^{3}$ we obtained earlier. Suppose we want to differentiate this result.

- Select the output $\frac{1}{3} x^{3}$ with the right mouse button (or hold down the option key when clicking for Mac users). A **Context** menu pops up.
- Select **Differentiate x**.

Immediately, *Maple* will automatically enter the input and show you the result:

```
R0 := diff(1/3*x^3,x);
```
$$R0 := x^{2}$$

(The name **R0** is automatically generated by *Maple* for your reference.)

Other Uses of Palettes

Greek Letters

In earlier versions of *Maple*, you had to spell out the name of a Greek letter to write Greek! For example, you need to define a variable named **alpha**, in order to see the Greek character α directly in front of you. Now, however, you can click the α button in the **Symbol** palette; then *Maple* will enter the word **alpha** automatically for you at the current location of the cursor.

For example, you can input an expression that assigns 3 to the variable α by clicking the α button, then typing ":=" and "**3**". *Maple* will automatically type:

```
alpha := 3;
```
$$\alpha := 3$$

Now you can click the [α] button anywhere you'd use the variable α, such as with:

```
4*[α] + 1;
```
$$13$$

This is particularly helpful, say, when you need to enter spherical coordinates. (The output is not shown here, however.)

```
plot3d([cos([θ])sin([φ]),sin([θ])sin([φ]),cos([φ])],
       [θ] = 0..2*[π], [φ] = 0..[π]);
```

Entering Math Expressions in Text Groups

You can use the **Expression** and **Symbol** palettes to enter mathematical expressions into the text of a worksheet. Suppose the cursor is located in a text region (please refer to Appendix A for information about text regions) and you want to add the comment:

"We need to calculate $\dfrac{\partial}{\partial x} 2xy^2$ to justify the answer."

- Press the [T] button in the tool bar to change to the text mode, then enter the first four words as usual.

- In the **Insert** menu choose **Standard Math.** Enter the formula $\dfrac{\partial}{\partial x} 2xy^2$ by choosing the [∂a/∂b], [a·b], and [aᵇ] buttons and typing the appropriate variables in the equation with the help of the *tab* key.

- Press the [T] button in the tool bar to change back to the text mode and continue typing the remaining sentence.

APPENDIX C

Create Your Own *Maple* Package

The Maple .m file

If you have written *Maple* routines that will be used regularly, it is very easy to put them in your own package. You can then recall all of these routines easily and quickly using the **with** command.

For example, suppose that you have created two routines named **draw1** and **leftHandSum**. The first, **draw1**, draws the graph of a function together with its derivative. The other, **leftHandSum**, computes the left hand sum (in integration) of a function over a given interval using a specified number of divisions.

To save these routines in a package named **myPackage1**, start a new *Maple* session and then enter the definitions of these routines. (We will improve these routines later.)

```
myPackage1[draw1] := proc( f, range)
    plot( [f, diff(f, x)], range,
    legend=["function","derivative function"]);
end proc:

myPackage1[leftHandSum] := proc( f, a, b, n)
    local k;
    k := (b-a)/n:
    evalf(add( subs(x= a+i*k, f), i=0..n-1)*k);
end proc:
```

Now, type:

```
save myPackage1, `myPackage1.m`;
```

You should now quit *Maple*. Next, find the **myPackage1.m** file inside the *Maple* directory or folder. Move that file to the **lib** directory or folder.

After you restart *Maple,* the **myPackage1** is ready for use. You can try it:

```
with(myPackage1);
```

    [draw1, leftHandSum]

```
draw1( x^2, x=-2..2 );
```

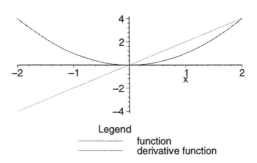

Legend
----------------- function
----------------- derivative function

```
leftHandSum( x^2, 1, 3, 10);
```

> 7.880000000

If you want to share a newly created *Maple* package across different computer systems or platforms, you'll find it more reliable to use a text file than the **.m** file. In the example above, you should create a text file, say **myPackage1.txt**, that contains the routines for **draw1** and **leftHandSum**. (This text file can be accessed by any computer system.) Another *Maple* user can store this text file in the *Maple* directory and type:

```
read `myPackage1.txt`;
```

He or she can then execute the two routines, and enter

```
save myPackage1, `myPackage1.m`;
```

to recreate the **.m** file for the package.

> **Note**. Beginning with *Maple* 8, you can also distribute your own *Maple* code using the multiple library support. For more information see **?repository, management**.

Design Suggestions for Packages

Design routines that can Accept General Input

If you want to publish a package that will be used by a lot of people, it is better to design routines that can accept general input.

For example, the **draw1** routine above only works for functions with variable x, because of the command **diff(f, x)**. One can rewrite the routine as:

```
myPackage1[draw1] := proc( f, inputRange)
plot( [f, diff(f, lhs(inputRange))], inputRange,
legend=["function","derivative function"]);
end proc:
```

The command **op(range)[1]** will extract the variable from the input **range**. As a result, the routine can now accept functions with different variables. For example,

```
draw1( cos(t), t=-2*Pi..2*Pi);
draw1( exp(1.2*s), s=-2..2);
```

Design syntax that's consistent with Maple

In the `leftHandSum` routine above, a user must enter the interval as **a, b**. This is not the standard way to enter a range in *Maple* and it can easily cause confusion. If you want to allow users to work with the standard entry format of **x = a..b**, then use

`lhs(x=a..b)` to extract the variable *x*.

`lhs(rhs(x=a..b))` to extract the number *a*.

`rhs(rhs(x=a..b))` to extract the number *b*.

As a result, we can rewrite the `leftHandSum` routine as

```
myPackage1[leftHandSum] := proc( f, inputRange, n )
    local a, b, k;
    a := lhs(rhs(inputRange)):
    b := rhs(rhs(inputRange)):
    k := (b-a)/n:
    evalf(add( subs(lhs(inputRange)= a+i*k, f ),
                    i=0..n-1)*k);
end proc:
```

Now, we can use it in the form:

```
leftHandSum(x^2, x=1..3,10);
leftHandSum(t^2, t=1..3,10);
```

Think about User Options

Sometimes you may want your routine to accept more options in the input. For example, in the `draw1` routine, we may want to allow the users to enter other plot options to control the output picture. These can be done with the `args[i]` and `nargs` commands which refer to the *i-th* argument and the total number of arguments respectively. We can rewrite the `draw1` routine as

```
myPackage1[draw1] := proc()
  local f, inputRange, n;
  f := args[1]:
  inputRange := args[2]:
  n := nargs:
  plot( [f, diff(f, lhs(inputRange))], inputRange,
      legend=["function","derivative function"],
      args[3..n]);
end proc:
```

It can now accept inputs such as:

```
draw1(x^2, x =- 2..2);        #without options
draw1(x^2, x=-2..2,
    thickness=2, scaling=constrained);  #with two extra options
```

Learning Calculus with *Maple*

General Public Package

As illustrated in many examples of this book, *Maple* can be an effective tool in learning Calculus or other math courses. There are many educational *Maple* packages that are currently available to the public. Instructors may find these packages very helpful. The largest collection of these packages is in the *Maple* Application Center, which can be found at:

http://www.mapleapps.com

The applications include single worksheets, packages, and collections of worksheets organized into courses.

The Student[Calculus1] package

Starting with *Maple* 8, there is a **Student[Calculus1]** package that contains many commands helpful in learning Calculus. Some of these commands can provide animations or graphs that demonstrate a concept of Calculus. Other commands can allow "Single-Step computation" to help students practice applying the limit, differentiation, and integration rules step-by-step.

Here we will give a quick summary on some of those visualization programs inside the **Student[Calculus1]** package. You should first load the package:

```
with(Student[Calculus1]);
```

Please note that Maplets have been developed to provide more convenient interfaces to the **Student[Calculus1]** package.

Visualization Commands for Learning About Derivatives

Example	Comment
`Tangent( sin(x), x=1, output = line);` `Tangent( sin(x), x=1, output = plot);`	"**output = line**" will give the equation of tangent line at the given point while "**output = plot**" will draw the graph of both the function and tangent line.

`increment := [0.5, 0.1, 0.01, 0.001];`
`NewtonQuotient( sin(x), x=1, output =`
`value, h = increment);`
 	
`NewtonQuotient( sin(x), x=1, output =`
`animation, h = increment);`
	"**output = value**" will compute the values of Newton Quotient at the given increment, while "**output = animation**" will illustrate graphically how the secant lines at the given increment points converge to the tangent line.

`RollesTheorem( sin(x), x=1..1+2*Pi,`
`output = points);`
 	
`MeanValueTheorem( sin(x), x=0..3,`
`output = plot);`
	"**output = points**" will compute the value(s) of the point(s) that satisfies the specified theorem at the given interval, while "**output = plot**" will illustrate graphically the specified theorem at the given interval.

Visualization Commands for Learning About Applications of the Derivative

Example	*Comment*
`DerivativePlot( (x-5)*exp(x), x = -3..3, order = 1..5);`	The given example will sketch the graphs of the given function and its first 5 derivatives in a single picture for x between -3 and 3.
`FunctionChart( x^3-9*x^2-48*x+52, x = -6..14);`	The **FunctionChart** will indicate in the picture, the regions where the function is increasing/decreasing, and concave up/down. Also it will indicate the roots, local max/min points and inflection points of the given function.

`TaylorApproximation( sin(x), x=1,`
`order = 3);`
 	
`TaylorApproximation( sin(x), x=1,`
`output = animation, order = 1..10);`
	The first example will give the Taylor approximation of order 3 at the point $x = 1$.
	
	With "**output = animation, order = 1..10**", we can see how the approximation is improved as the order is increased from 1 to 10.

`NewtonsMethod( sin(x), x=2, output =`
`sequence);`
 	
`NewtonsMethod( sin(x), x =2, output =`
`plot);`
	"**output = sequence**" will generate a sequence of points from the Newton's method to approximate the zero of a function starting at the given point, while "**output = plot**" will illustrate graphically how these points are selected from the Newton's method.

Visualization Commands for Learning About Integrals

Example	*Comment*

`AntiderivativePlot( x^3 +2*x^2 -1,`
`x=-2..2);`
 	
`AntiderivativePlot( x^3 +2*x^2 -1,`
`x=-2..2, showclass);`
	The first command will draw the given function together with one of its antiderivatives. When we add the **showclass** option in the second command, *Maple* will draw a family of antiderivatives.

`ApproximateInt( x^2, x=0..4, method = midpoint, output= plot);`	It will sketch the graph of the given function (in the specified interval) together with the rectangles that approximated the definite integral.
	Options for **method** include **left, right, trapezoid, lower, upper, random** and nonrectangular rules like **simpson** and **newtoncotes**.
	If we use **output = animation** instead, then the picture will demonstrate that as the number of rectangles increased, the definite integral is better approximated.

Visualization Commands for Learning About Applications of Integration

Example	*Comment*
`FunctionAverage(sin(x), x=0..Pi, output = plot);`	It will sketch the graph of the given function and its average value in the specified interval.

`VolumeOfRevolution( sin(x)+2, x = 0..Pi, output = plot);`
 	
`SurfaceOfRevolution( sin(x)+2, x = 0..Pi, output = plot);`
	These commands will draw the object corresponding to the volume or surface of revolution of the given function along the horizontal axis. If we want to rotate the function along the vertical axis, we need to include the option, **axis = vertical**.

Single Step Computation for Differentiaion

Students can use the single-step routines inside the **Student[Calculus1]** package to learn more about the technique of differentiation and integration. By using the **Rule** command, one can specify which differentiation or integration rule to use in each step. For example,

```
with(Student[Calculus1]) ;
Diff(x^3+x*sin(x^2), x);
```

$$\frac{\partial}{\partial x}(x^3 + x\sin(x^2))$$

`Rule[sum](%);`   # Apply the Sum rule of differentiation.

$$\frac{\partial}{\partial x}(x^3 + x\sin(x^2)) = \left(\frac{\partial}{\partial x}x^3\right) + \left(\frac{\partial}{\partial x}x\sin(x^2)\right)$$

`Rule[power](%);`   #Apply the power rule of differentiation

$$\frac{\partial}{\partial x}(x^3 + x\sin(x^2)) = 3x^2 + \left(\frac{\partial}{\partial x}x\sin(x^2)\right)$$

`Rule[product](%);`   #Apply the product rule

$$\frac{\partial}{\partial x}(x^3 + x\sin(x^2)) = 3x^2 + \left(\frac{\partial}{\partial x}x\right)\sin(x^2) + x\left(\frac{\partial}{\partial x}\sin(x^2)\right)$$

`Rule[power](%);`   # or use `Rule[Identity](%);`

$$\frac{\partial}{\partial x}(x^3 + x\sin(x^2)) = 3x^2 + \sin(x^2) + x\left(\frac{\partial}{\partial x}\sin(x^2)\right)$$

`Rule[chain](%);`   #Apply the chain rule

$$\frac{\partial}{\partial x}(x^3 + x\sin(x^2)) = 3x^2 + \sin(x^2) + x\left(\frac{\partial}{\partial\_X}\sin(\_X)\right)\Bigg|_{\_X\,=\,x^2}\left(\frac{\partial}{\partial x}x^2\right)$$

Rule[sin](%);   #Use the differentiation rule for the sine function

$$\frac{\partial}{\partial x}(x^3 + x\sin(x^2)) = 3x^2 + \sin(x^2) + x\cos(x^2)\left(\frac{\partial}{\partial x}x^2\right)$$

Rule[power](%);

$$\frac{\partial}{\partial x}(x^3 + x\sin(x^2)) = 3x^2 + \sin(x^2) + x\cos(x^2)(2x)$$

Single Step Computation for Integration

Another example for integration:

Int(x^3+x*cos(2x), x);

$$\int\ x^3 + x\cos(2x)\ dx$$

Rule[sum](%);   # Apply the Sum rule of integration

$$\int\ x^3 + x\cos(2x)\ dx = \int\ x^3 dx + \int\ x\cos(2x)\ dx$$

Rule[power](%);   # Apply the Power rule of integration

$$\int\ x^3 + x\cos(2x)\ dx = \frac{x^4}{4} + \int\ x\cos(2x)\ dx$$

Rule[change, u=2*x](%);   # Use integration by substitution $u = 2x$.

$$\int\ x^3 + x\cos(2x)\ dx = \frac{x^4}{4} + \int\frac{1}{4}\cos(u)\ u\ du$$

Rule[constantmultiple](%);   # Pull out the constant factor.

$$\int\ x^3 + x\cos(2x)\ dx = \frac{x^4}{4} + \frac{1}{4}\int\ \cos(u)\ u\ du$$

Rule[parts,u,sin(u)](%);   # integration by parts. The user must specify the parts.

$$\int\ x^3 + x\cos(2x)\ dx = \frac{x^4}{4} + \frac{1}{4}u\sin(u) - \frac{1}{4}\int\ \sin(u)\ du$$

Rule[sin](%);   # Use integration rule for sine function.

$$\int\ x^3 + x\cos(2x)\ dx = \frac{x^4}{4} + \frac{1}{4}u\sin(u) + \frac{1}{4}\cos(u)$$

Rule[revert](%);   # Undo the earlier substitution.

$$\int\ x^3 + x\cos(2x)\ dx = \frac{x^4}{4} + \frac{1}{2}x\sin(2x) + \frac{1}{4}\cos(2x)$$

In general the rules that one can use for differentiation in *Maple* include:

sum, difference, product, quotient, constant, constantmultiple, power, identity, chain, and **Int,** as well as the name of any function of one variable recognized by *Maple*.

The rules for integration in *Maple* include:

sum, difference, product, quotient, constant, constantmultiple, power, Diff, parts, change, revert, partialfractions, rewrite, join, split, flip, as well as the name of any function of one variable recognized by *Maple*.

Note that the **parts**, **change**, **rewrite**, and **split** rules require the user to provide information on how to apply those rules.

Undo and Hint   In case that you make a mistake, you can **Undo** the process. If you are unsure which rule to use in the next step, you ask for a **Hint**. (If you ask for a hint, it is worthwhile to change the **infolevel** to **1** to get an expanded description of the hint.)

> **Int(sqrt(1-x^2), x);**

$$\int \sqrt{1-x^2}\, dx$$

A common mistake for the student is to substitute $u = 1 - x^2$. Let us see what will happen:

> **Rule[change u = 1-x^2](%);**

$$\int \sqrt{1-x^2}\, dx = \int -\frac{\sqrt{u}}{2\sqrt{-u+1}}\, du$$

This does not look right, we have to undo it.

> **Undo(%);**

$$\int \sqrt{1-x^2}\, dx$$

We can ask *Maple* for a hint:

> **infolevel[Student]:= 1;**
> **Hint(%);**

Hints:

1. Integrals involving expressions of the form sqrt(a^2-x^2) can often be simplified using the substitution x=a*sin(u).
2. Use the substitution −1+1/x^2 = u^2.
3. Integrals involving expressions of the form sqrt(a^2-x^2) can often be simplified using the substitution u = sqrt((a-x)/(a+x)).

$$\left[change, x = \sin(u), u \right], \left[change, -1 + \frac{1}{x^2} = u^2, u \right], \left[change, u = \sqrt{\frac{1-x}{1+x}}, u \right]$$

Now, we can continue the integration with:

> **Rule[change, x=sin(u)](%%)**   #Note that we use **%%** instead of **%** here.

$$\int \sqrt{1-x^2}\, dx = u - \int \sin(u)^2\, du$$

Note. In *Maple* 9, you can use the **Student[Calculus1]** package with the Maplets technology. This provides a graphical interface for users. Visit www.mapleapps.com for more information.

APPENDIX E
Scientific Constants

The `ScientificConstants` Package

Physical Constants

In version 8, *Maple* introduced the **ScientificConstants** package that gives users access to many constants used in science. There are about 100 physical constants contained in the package. You can see the whole collection with:

```
with(ScientificConstants):
GetConstants('names');
```

Avogadro_constant, Bohr_magneton, Bohr_radius, Boltmann_constant, Compton_wavelength, Faraday_constan, Hartree_energy, Josephson_constant, Loschmidt_constant, Newtonian_constants_of_gravitation, Planck_constant, Plank_length, Planck_mass, Rydberg_constant, Stefan_Boltzmann_constant, Thomson_cross_section, Wien_displacement_law_constant, alpha_particle_mass, atomic_mass_constant, characteristic_impedance_of_vacuum, classical_electron_radius, conductance_quantum, deuteron_magnetic_moment, deuteron_mass, electron_g_factor, electron_gyromagnetic_ratio, electron_magnetic_moment, electron_magnetic_moment_anomaly, electron_mass, elementary_charge, fine_structure_constant, first_radiation_constant, first_radiation_constant_for_spectral_radiance, helion_mass, magnetic_flux_quantum, mass_of_Earth, mass_of_Sun, molar_gas_constant, molar_volume_of_ideal_gas, muon_Compton_wavelength, muon_g_factor, muon_magnetic_moment_anomaly, muon_mass, neutron_Compton_wavelength, neutron_g_factor, neutron_gyromagnetic_ratio, neutron_magnetic_moment, neutron_mass, nuclear_magneton, permeability_of_vacuum, permittivity_of_vacuum, proton_Compton_wavelength, proton_g_factor, proton_gyromagnetic_ratio, proton_magnetic_moment, proton_magnetic_shielding_correction, proton_mass, radius_of_Earth, second_radiation_constant, shielded_helion_gyromagnetic_ratio, shielded_helion_magnetic_moment, shielded_proton_gyromagnetic_ratio,shielded_proton_magnetic_moment, speed_of_light_in_vacuum, standard_acceleration_of_gravity, tau_Compton_wavelength, tau_mass, von_Klizing_constant.

or we can see the symbols of constants with:

```
GetConstants();
```

$E_h, F, G, G_O, K_J, M_{Earth}$, ...   #We will not show the complete result here.

To find out about the gravitational constant, for example, we will enter:

```
GetConstant(G);    # or
GetConstant(Newtonian_constants_of_gravitation);
```

Newtonian_constants_of_gravitation, symbol =G, value=..6673 10^{-10},

unvertanity= .10 10^{-12} *, units* = $\dfrac{m^3}{kg\ s^2}$

To use the value or unit of the gravitational constant for computation, we enter:

```
GetValue(Constant(G));
```

$$.6673 \ 10^{-10}$$

```
GetUnit(Constant(G));
```

$$\left[\frac{m^3}{kg \ s^2}\right]$$

Properties of Elements

We can also find out the properties of each chemical element from the **ScientificConstants** package.

For example, to find the information about oxygen, we can enter:

```
GetElement(Oxygen);     # you can use the element's name,
GetElement(O);          # or you simply use its symbol
```

8, *symbol* = O, *name* = oxygen, *names* ={*oxygen*},

electronaffinity = [*value* = 1.4611103, *uncertanity* = .7 10^{-6}, *units* = *eV*],
density = [*value* = (**proc**() ... **end proc**), *uncertainty* = *undefined*, *units* = *L*],
ionizationenergy = [*value* = 13.6181, *uncertainty* = *undefined*, *units* =*eV*],
atomicweight = [*value* = 15.9994, *uncertainty* = .0003, *units* = *u*],
boilingpoint = [*value* = 90.20, *uncertainty* = *undefined*, *units* = *K*],
electronegativity = [*value* = 3.44, *uncertainty* = *undefined*, *units* = 1],
meltingpoint = [*value* = 54.36, *uncertainty* = *undefined*, *units* = *K*]

To find the ionization energy of oxygen and use it in the *Maple* computation, we type:

```
GetValue(Element(Oxygen,ionizationenergy));
```

13.6181

There are other commands such as **GetIsotopes** and **GetUnit** that can give you more information about an element. You can find out more about these commands from *Maple*'s **Help** menu.

> **Note**. *Maple* 9 contains a **ScientificErrorAnalysis** package to complement **ScientificConstants**. This new package is available for download for *Maple* 8 from www.MaplePrimes.com.

Index